Companion Guidebook
to Project Management for Mining

by Terry Owen, Russell Owen,
and Kristin Rygnestad

PUBLISHED BY THE
SOCIETY FOR MINING, METALLURGY & EXPLORATION

Society for Mining, Metallurgy & Exploration (SME)
12999 E. Adam Aircraft Circle
Englewood, Colorado, USA 80112
(303) 948-4200 / (800) 763-3132
www.smenet.org

SME advances the worldwide mining and underground construction community through information exchange and professional development. SME is the world's largest association of mining and minerals professionals.

Information contained in this work has been obtained by SME from sources believed to be reliable. However, neither SME nor its authors and editors guarantee the accuracy or completeness of any information published herein, and neither SME nor its authors and editors shall be responsible for any errors, omissions, or damages arising out of use of this information. This work is published with the understanding that SME and its authors and editors are supplying information but are not attempting to render engineering or other professional services. Any statement or views presented herein are those of individual authors and editors and are not necessarily those of SME. The mention of trade names for commercial products does not imply the approval or endorsement of SME.

ISBN 978-0-87335-015-0
Ebook 978-0-87335-021-1

Contents

Preface

Developing an advanced mineral resource into an operating mine-production facility has become increasingly complex with timelines stretched so dramatically that many companies can now advance a new project only once a decade. At the same time, global demand for minerals continues to surge, intensifying the pressure on the mining industry to identify and develop new resources. As a result, more mining professionals—often for the first time—are being tasked with guiding a mineral resource through the demanding stages of project development.

Project Management for Mining Companion Guidebook is designed to be a critical resource for those professionals. It delivers the focus, structure, and insight needed to navigate the exciting—and frequently challenging—process of transforming a mineral resource into a successful operating mine. As an extension of *Project Management for Mining 2nd edition* authored by Robin Hickson and Terry Owen, it serves as a concise, at-hand reference for anyone responsible for leading a mining project.

The SME *Project Management for Mining 2nd edition* offers a holistic view of developing a mining project. Authors Robin Hickson and Terry Owen distilled mining-industry best practices drawn from Owners, EPCMs, OEMs, and contractors. Capturing the collective wisdom of seasoned project leaders from around the world, the Handbook provides a comprehensive foundation for successful mining-project delivery.

Project Management for Mining Companion Guidebook, authored by Terry Owen and Russell Owen, presents a focused set of guidelines that distill the 780-page handbook into the critical tasks within each project stage and phase. It sharpens attention on the opportunity-through-viability stages essential in today's dynamic environment. With integrated figures and checklists from the *Project Management for Mining 2nd edition*, it functions as a clear, concise guide for anyone charged with leading a mining project. Together, the guidebook and the handbook define best practices for major or mega greenfield projects as well as brownfield projects of any size within operating facilities.

The *Project Management for Mining 2nd edition* has sold more than 2,200 copies and is used by mineral organizations in approximately 75 countries. It has also been adopted as a teaching aid by 46 learning institutions across 22 countries.

The Project Management for Mining course, taught by the authors since 2015, 137 organizations from 47 countries have sent more than 500 personnel, representing 127 different minerals-industry titles, to complete the program. Insights gained from these ten years of instruction have directly added to and refined the focus in *Project Management for Mining Companion Guidebook.*

Designed for international use, the guidebook features a side-by-side English/Spanish format. Kristin Rygnestad's leadership in managing the complexity of translation now offers Spanish-speaking professionals a clear and accessible pathway into the full value of the *Project Management for Mining 2nd edition.* This bilingual structure strengthens global adoption by making essential project-management principles easier to navigate and apply across a bilingual profession.

Acknowledgments

The journey Robin and Terry traveled to produce the two editions of the SME *Project Management for Mining* was described in the Acknowledgement of the 2nd edition as follows:

"Our combined experiences in the global minerals industry are the central source for the effective project management tactics and methods that we seek to convey within this book.... This publication is a culmination of the wisdom we gleaned and gathered from so many others as we worked our way through almost 500 projects following our entry into the mining profession." It took four years to produce the first edition and another two years to complete the second".

The development of *Project Management for Mining Companion Guidebook* has followed a different journey—one built upon the best-practices foundation established in *Project Management for Mining*. That foundation includes the only book of its kind and presents a best-practices, holistic approach to complex mining projects. More than 2,200 copies have been sold, and it is known to be used by mineral organizations in roughly 75 countries. It has also been adopted as a teaching aid by 46 learning institutions in 22 countries.

As the handbook reached a broader audience, demand grew across Spanish-speaking regions of the Americas for access to these best practices in Spanish. This need inspired a collaborative effort among mining professionals in North America and South America. Using key elements from the *Project Management for Mining*, materials from the *Project Management for Mining* course, emerging content related to mining projects, and direct requests from Spanish-speaking professionals, this collaboration produced the *Project Management for Mining Companion Guidebook* in both English and Spanish.

Several key individuals dedicated countless hours to the success of this effort. These include Russell Owen, Kristin Rygnestad, Renzo Macher, and many of their colleagues and contacts.

Renzo Macher, a Peruvian national with international experience, currently the Vice President of Projects and Innovation for Compañía de Minas Buenaventura, has been a strong advocate for this guidebook and a key collaborator. He generously reviewed the final Spanish-language version for accuracy. Two additional individuals provided valuable contributions: Additional Spanish speaking contacts providing input to the translation include: Alvaro Marino from Columbia with experience in Colombia, Panama and Dominican Republic; Patricia Suegart from Venezuela with experience in Venezuela and Dominican Republic; Ricardo Huancaya and Denise Chavez from Peru. Bruce Marsh, recognized for his mastery of the complexities of international environmental and social engagements in the mining industry, and Chris McGoldrick, whose project development and construction expertise consistently anticipates what is possible for efficiency in project execution.

We are extremely grateful to the two South American mining companies that provided funding and key resources for this project: Compañía de Minas Buenaventura S.A.A. in Peru—along with several other Peruvian mining professionals—and Minera San Cristóbal S.A. Bolivia, both of which contributed significant support.

A special thanks goes to Robin Hickson, who led the development of the first *Project Management for Mining* Handbook. Robin provided an in-depth peer review of this guidebook and offered invaluable insights that only he can provide. Robin and Terry have shared many experiences across the decades in the international mining industry.

Finally, we extend our appreciation to Melissa Serdinsky, under the banner of the Society of Mining, Metallurgy and Exploration (SME), for encouraging this project and helping bring it to life.

Our collective hope is that this English–Spanish guidebook will empower the Spanish-speaking mining community and, together with *Project Management for Mining, 2nd ed* foster more successful mining projects.

About the Authors

Terry Owen was born in the northwest of the USA. During his college years Terry served as an US Army helicopter pilot in Vietnam prior to earning his Bachelor of Science degree as a mining engineer. During his career he obtained a degree from the Harvard Business School PMD program and is a QP. He has over 48 years of international experience in the minerals industry developing and operating mines in multicultural environments around the globe on six continents. He has served in various executive management operating and project development roles with special expertise developing complex, difficult, remote projects in greenfield and brownfield environments, from initial definition through all stages of development including advanced exploration, feasibility, engineering, construction, operations team development and readiness, commissioning, start- up and optimization of operations. He has worked in Gold, Copper, Lead, Silver, Zinc, Molybdenum, Iron ore and Coal. Through a life of hard work, resilient problem solving, success working with multiple cultures wrapped up in an unexpected, challenging, rewarding mining career that included having successfully executed billions of dollars of project and operations start-ups in ten different countries in Africa, Australasia, South America, North America, Russia, from locales in the frozen Arctic to the rainless deserts of Chile; from the 4800m (14,750ft) high altitudes of Bolivia, Peru, West Papua (Island of New Guinea) to the coastal swamps of Indonesia and Papua New Guinea. He currently consults with mining industry organizations and boards relating to project delivery and coaches/advises development of projects for companies. He is the co-author of SME's *Project Management for Mining,* 1st and 2nd editions and has instructed the course Project Management for Mining at the Colorado School of Mines and with companies. To know more about Terry and his mining adventure read his memoir *Unexpected Paths.*

Russell Owen 20 years of international operations and project experience in the Americas, Democratic Republic of Congo, Pakistan, and Caribbean. He has served in a variety of management positions as well as board and joint venture committees. Russell specializes in mining and processing optimization through integrated mineral resource management and has extensive experience with modern modeling practices in geology, geometallurgy, geotechnical and hydrogeology. He is experienced in difficult environments, working with local team and communities to achieve social acceptance to operate. He has worked on optimizing existing mega-projects globally and drove many of the technical aspects of the Reko Diq Copper and Gold feasibility study in far western Pakistan. Russell has experience in base metals, precious metals, and commercial minerals. He is a Fellow with AusIMM and has served as Subject Matter Expert and QP.

Kristin Rygnestad brings over 29 years of experience in project controls and financial analysis within the mining industry. Originally from Norway, she pursued mining engineering studies in the United Kingdom. After gaining professional experience in Chile, she relocated to the United States to earn a master's degree in mineral economics from the Colorado School of Mines.

Kristin possesses deep industry knowledge, technical expertise, and strong leadership skills. She has held senior management roles in North and South America and has served as a consultant for international mining companies with projects in Romania, Chile, Peru, Bolivia, Mexico, Colombia, Argentina, Vietnam, Eritrea, Tanzania, Canada, and the United States.

She has worked on projects through all phases, from identification and planning to execution and closure. Her expertise includes project controls setup and implementation, developing project controls procedures, cost estimating from an owner's perspective, project reporting, and managing interfaces between project controls and the owner's finance and accounting departments. Kristin also has extensive experience in financial analysis to support management decisions on mergers and acquisitions, strategic planning, investments in greenfield and brownfield projects, operational budgeting, and project financing.

Kristin is fully bilingual in English and Spanish and maintains her native Norwegian despite few opportunities to use it.

SECTION 1

Purpose and Scope

PURPOSE

The *Project Management for Mining Companion Guidebook (CG)* outlines a clear and concise roadmap for managing complex mining projects, as detailed in *Project Management for Mining, 2nd ed.* (*PMfM*). This roadmap is designed to support the appointed project advocate and/or project manager (PM) through every stage of project development, ultimately leading toward the successful execution of mining projects.

The *CG* provides Spanish-speaking readers with guidance on where to find the appropriate detail in *PMfM*. Together, the two books present a best-practice process for steering project development from idea, through senior management approval, to on-site execution.

PMfM has been written to help miners succeed in the exceedingly challenging world of project management. Mining projects are complex, and the path to successful project completion is embedded with situational uncertainties and strewn with obstacles.

PMfM provides the originator of a mining idea or mineral opportunity with the necessary guidance to develop that idea or opportunity into a well-defined project that will successfully meet the corporation's business objectives.

PMfM identifies the roles, responsibilities, and accountabilities of the individuals involved in the development and execution of the project. It provides a skeleton for, and reference to, the various documents that compose the mining industry's soundest practices. It identifies the interrelationships between the parties and disciplines in a typical project. By following the procedures contained within, a mining organization can be assured that the project is properly characterized, project viability has been adequately challenged, the correct questions are being answered, and the appropriate process is being implemented.

PMfM is also a tool for the project team to draw from when facing unfamiliar problems, providing project staff with a more complete understanding of how to achieve project effectiveness.

Message for the Spanish Speaker: This guidebook is primarily intended for Spanish speakers with English as a second language. It provides a set of reference tools to help them navigate and locate key information within *PMfM*.

> **Note: Each *section* of *CG* is designed to direct the reader to the corresponding *chapter* in *PMfM*, which must be thoroughly understood for successful implementation. Therefore, *PMfM* must be used in conjunction with *CG*.**

COMPANION GUIDEBOOK SCOPE

CG provides a set of guidelines laid out in a parallel English/Spanish format. This format outlines a clear and concise roadmap for managing major mining greenfield projects and guides the reader to key points in *PMfM*. The roadmap guides the appointed project advocate and/or project manager through the project development process, covering all five stages of project development (first idea or opportunity, viability, planning, execution, and completion), and contains a set of focused flowcharts, checklists, and references found in *PMfM* in both Spanish and English. In response to recent project delays, cost overruns, and failures, the *CG* places greater emphasis on the opportunity-through-viability stage. The objective is to ensure every project is rigorously scoped, well structured, tightly controlled, and properly funded from the outset.

Note: (1) To provide clarity between *PMfM* and this *CG*, *PMfM* is divided into chapters and the *CG* is divided into sections. (2) All definitions and terminology are based on English-language terms defined in *PMfM*.

MAJOR GREENFIELD PROJECTS

The guidebook focuses on major greenfield project development, as detailed in *PMfM*.

- *CG*'s overall focus is on the key steps necessary to ensure the advancement of an opportunity through the project stages to completion.
- The *CG* is organized into five stages of project life (opportunity, viability, planning, execution, and completion). Each section has the following components:
  - The objective, outlining areas of focus for each section.
  - References to *PMfM* chapters, figures, flowcharts, and checklists.
  - Scope and key points for each section.
  - Selected figures, flowcharts, and checklists drawn directly from *PMfM*, with others modified or introduced in the *CG*.

BROWNFIELD PROJECTS

A brownfield project at an existing operation follows the same principles detailed in *PMfM*. The modifying assumption is that, for a mature, currently operating brownfield facility expansion, one would not need the same level of effort as for a new greenfield project. However, working within an existing operation requires more diligence in certain areas.

Should the project proponent be focused on a brownfield project, most components are the same, but some require specialized focus. See *CG* Section 18 and *PMfM* Chapter 38 for further detail.

SECTION 2

Overview

OBJECTIVE

To outline the necessary components of a project, it is necessary to understand the stages of project development, the stage gates for advancing through the stages of a project, the ethical issues surrounding projects, and the opportunity stage of projects as it relates to the rapidly developing mining landscape.

Each section of these guidelines is designed to direct the reader to the corresponding chapter in *Project Management for Mining* (*PMfM*), which must be thoroughly understood for successful implementation. *PMfM* must be used alongside these guidelines.

REFERENCES

Item	*PMfM*	Page
Introduction	Chapter 1	3
Ethics	Chapter 3	23
Governance	Chapter 4	33
Scoping Evaluation	Chapter 5	49

CG Figure	Figure Caption	*PMfM* Reference
Figure 2.1	Introduction (project life cycle)	Flowchart 1.1, p.18
Figure 2.2	Project stages timeline	Figure 1.3, p.11
Figure 2.3	Project management activities	Figure 1.1, p.6
Figure 2.4	Maximize probability of project success	New Figure
Figure 2.5	Governance program (stage gate)	Figure 4.4, p.42
Figure 2.6	Opportunity activities	Figure 5.1, p.52

Checklist Title	*PMfM* Reference
Utilizing guidebook	New Checklist

SCOPE

Components of Mining Projects

There are multiple stages to a project, from idea to completion. While each stage will have some overlap and is not entirely sequential, each stage builds upon the information and decisions reached in the preceding stage. While each project is different and the complexities will vary, all successful projects share the core components of project management.

The Introduction flowchart (Figure 2.1) ties together the various project phases described in *PMfM*.

- Each box is identified and referenced to the *PMfM* chapter that contains the details.
- The far-right column shows the project stages.
- The center column identifies the different phases and the order they follow.
- The upper-left column shows the stage -gating process, when it occurs, and the next step to be undertaken.

Project Stages and Their Natural Progression

There are five core stages in the life of all projects. They are defined as first idea or opportunity, viability, planning, execution, and completion. Figure 2.2 illustrates how these stages develop over a timeline.

By studying Figure 2.2, one can understand how certain components of the project delivery process must be undertaken early in the process and continue to be developed as the project progresses.

Project Management

What Is Project Management?

Project management is the process by which a team of people successfully guides a project using the elements of planning, analysis, directing, monitoring, problem solving, and communicating. They take an idea from the opportunity stage through development to the achievement of specific, established corporate objectives. They deliver these objectives within set scope, cost, schedule, and quality constraints.

The objective of *PMfM* and this *Companion Guidebook* (*CG)* is to bring the project management world into better focus, to place project managers in positions where they can anticipate rather than respond, and to lay out the project execution path in straightforward patterns with understandable success strategies that will achieve the desired outcome.

Figure 2.3 illustrates, in a flowchart format, how these various project management activities tie together (*CG* Section 3).

Project management is not a complex process. There are four basic steps to a successful project outcome:

1. Develop a definitive project scope and a project-specific execution plan.
2. Use qualified management personnel.
3. Create the project control mechanisms upfront (documents, tools, and procedures).
4. Control engineering, construction, and start-up activities during project execution.

Maximize Probability of Success While Reducing Project Risk

The results of mining projects in the first 2.5 decades of the 21st century show that major projects fail 85% of the time due to schedule extension, cost overrun, or failure to produce at the intended output and cost. Mega-project results are even worse, with such projects requiring 40% to 90% more capital and 12 to 18 additional months to complete.

In a major mining project, over 60% to 70% of the expenditure occurs during the execution stage of the project, and this is when the cost and schedule impacts occur.

The fundamental fact is that the failures of these projects are most often the result of an inadequate work effort to define and derisk the project (*CG* Section 4). The failure to do what best practice says is required during the viability stage has a significant impact on whether the execution stage performs as expected and the mine-plant operation is successful.

Advancing a project through the series of studies is about improving project definition and greatly reducing the risks to project success. This requires a much higher spend to advance the reliability of data models for mine, process, water, and engineering to 35% during the viability stage. This is illustrated in Figure 2.4.

Governance and Stage Gate Controls (CG Section 5)

A main component of project governance is providing a series of go/no-go decision points over the course of the project life cycle. Figure 2.5 outlines the governance function and its go/no-go decision gates related to the project management function. Stage gates are discussed in detail in *PMfM* Chapter 6.

Ethics

What makes mining projects so exciting is that each one is unusual and unique. But these projects can come with a downside. The project manager may encounter pressure, often internal, to spin, mischaracterize, or even misstate project information in a positive but unrepresentative manner to advance the project. *PMfM* Chapter 3 seeks to describe the conflicts, real or perceived, that may occur during the course of a project and to explore how project managers can best approach these issues from an ethics viewpoint. More importantly, *PMfM* Chapter 3 seeks to bolster the project manager's resolve to do the right thing.

Idea or Opportunity Stage (CG Section 6)

In addition to advanced exploration opportunities, at an operating mine there can be any number of opportunities that operators identify, including a new ore body, going underground from an open pit (or the opposite), process changes, major debottlenecking, an additional process circuit, a major tailings change, material-handling changes, etc.

Figure 2.6 shows the various elements one goes through from the first idea into opportunity development and scoping evaluation prior to knowing whether a project should move into the viability stage. The identification and evaluation of ideas is often a very fluid process that changes as the operators and other stakeholders examine the idea(s).

The important and critical point for this flowchart (Figure 2.6) is to understand that this idea and opportunity evaluation is often an effort completed by the operating organization and is not necessarily set to a project-related timeline or necessarily managed by an experienced PM.

However, once completed, reviewed by management, and approved to move forward into the viability stage, it is mandatory that the next steps are handled according to the project delivery process outlined in *PMfM* and this *CG*.

Identifying the Opportunity (Advanced Exploration or Idea)

There are many areas within a company from which a viable idea can originate. The next step is to expand and document the idea into a credible and reviewable opportunity.

- The first task is the development of the opportunity.
- The opportunity is then advanced to the viability stage.

Once the Scoping Evaluation has been completed, reviewed, and accepted by management, the project development process moves to the viability stage with very specific procedural steps and processes.

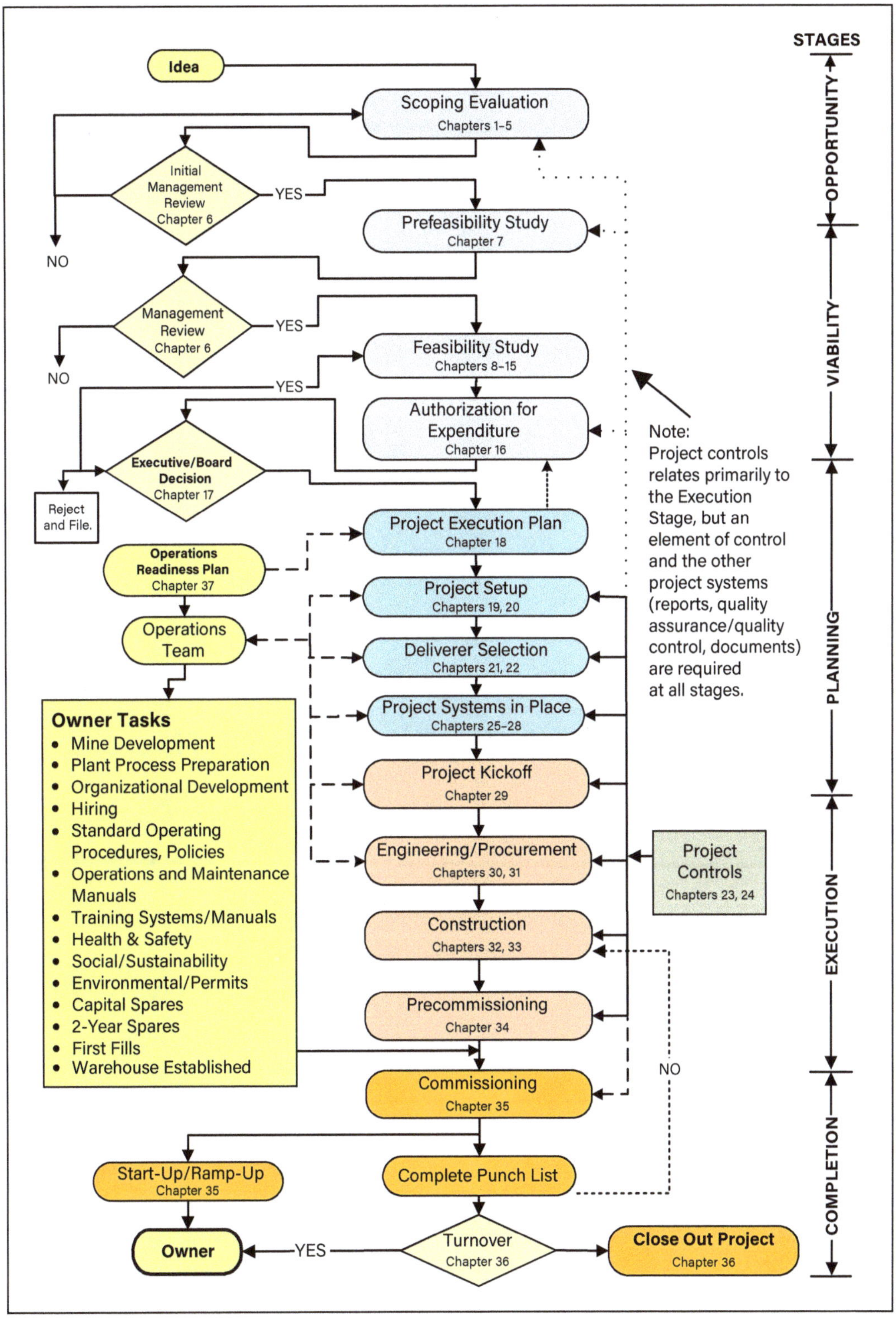

Figure 2.1: Introduction (project life cycle)

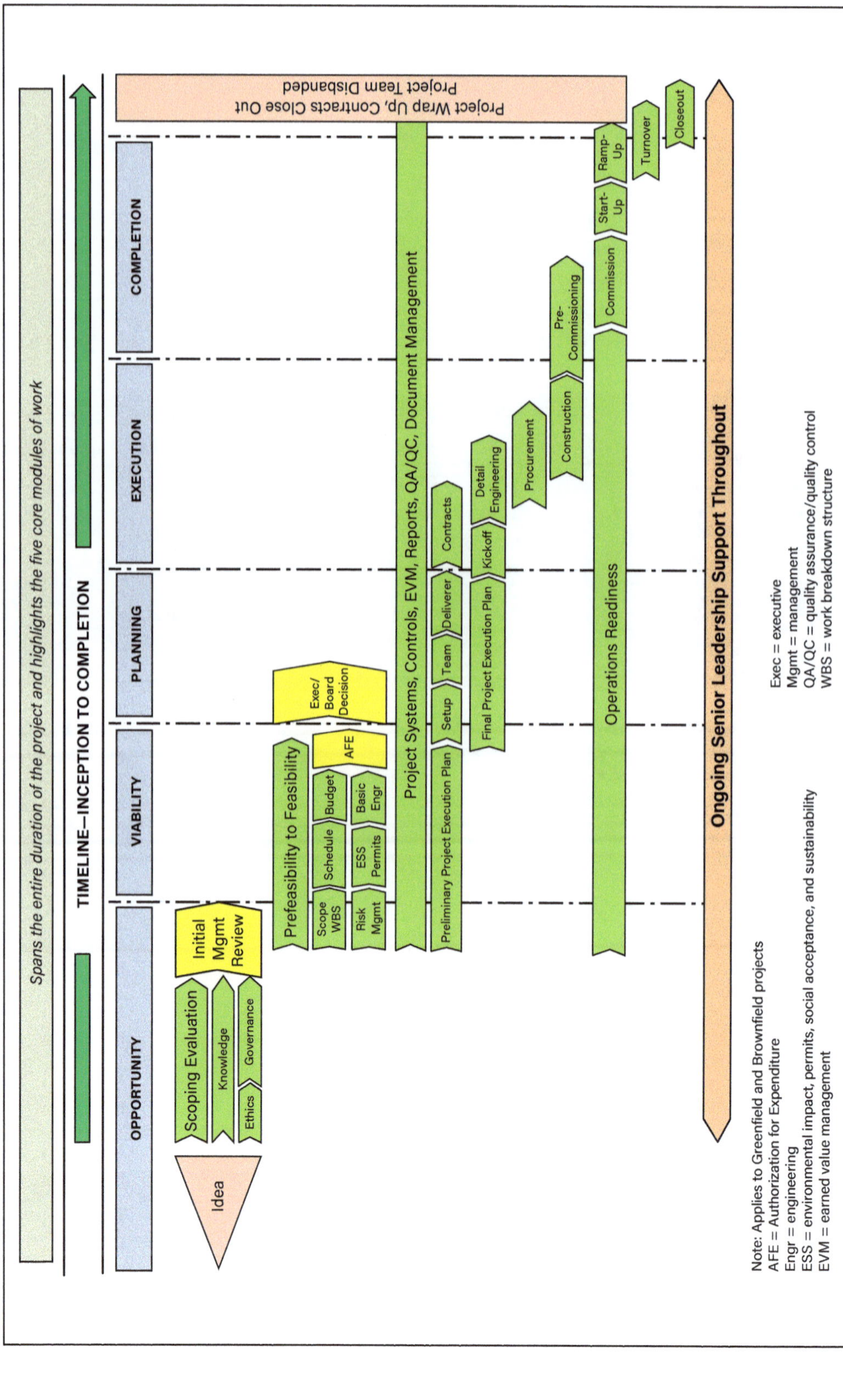

Note: Applies to Greenfield and Brownfield projects
AFE = Authorization for Expenditure
Engr = engineering
ESS = environmental impact, permits, social acceptance, and sustainability
EVM = earned value management

Exec = executive
Mgmt = management
QA/QC = quality assurance/quality control
WBS = work breakdown structure

Figure 2.2: Project stages timeline

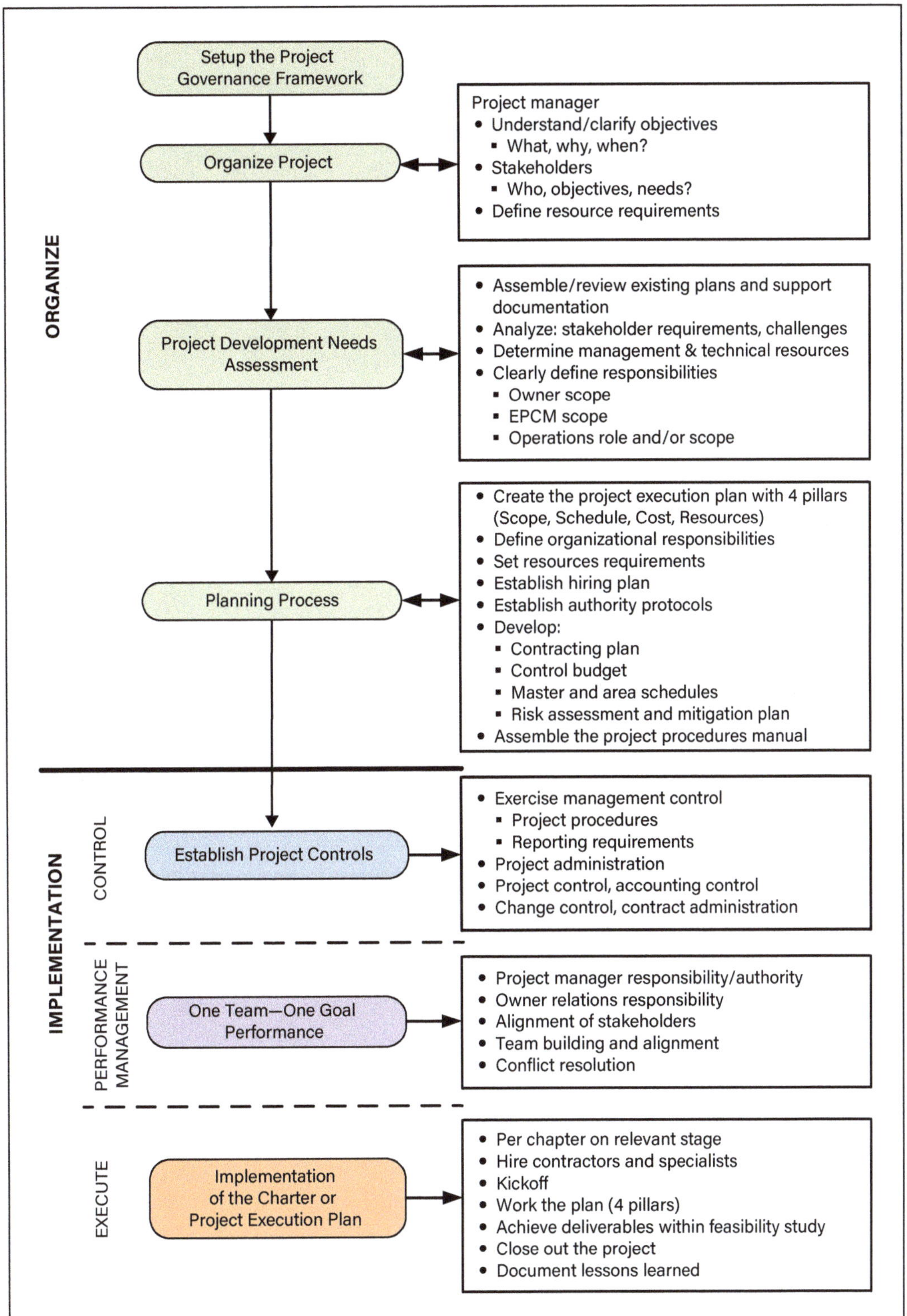

Figure 2.3: Project management activities

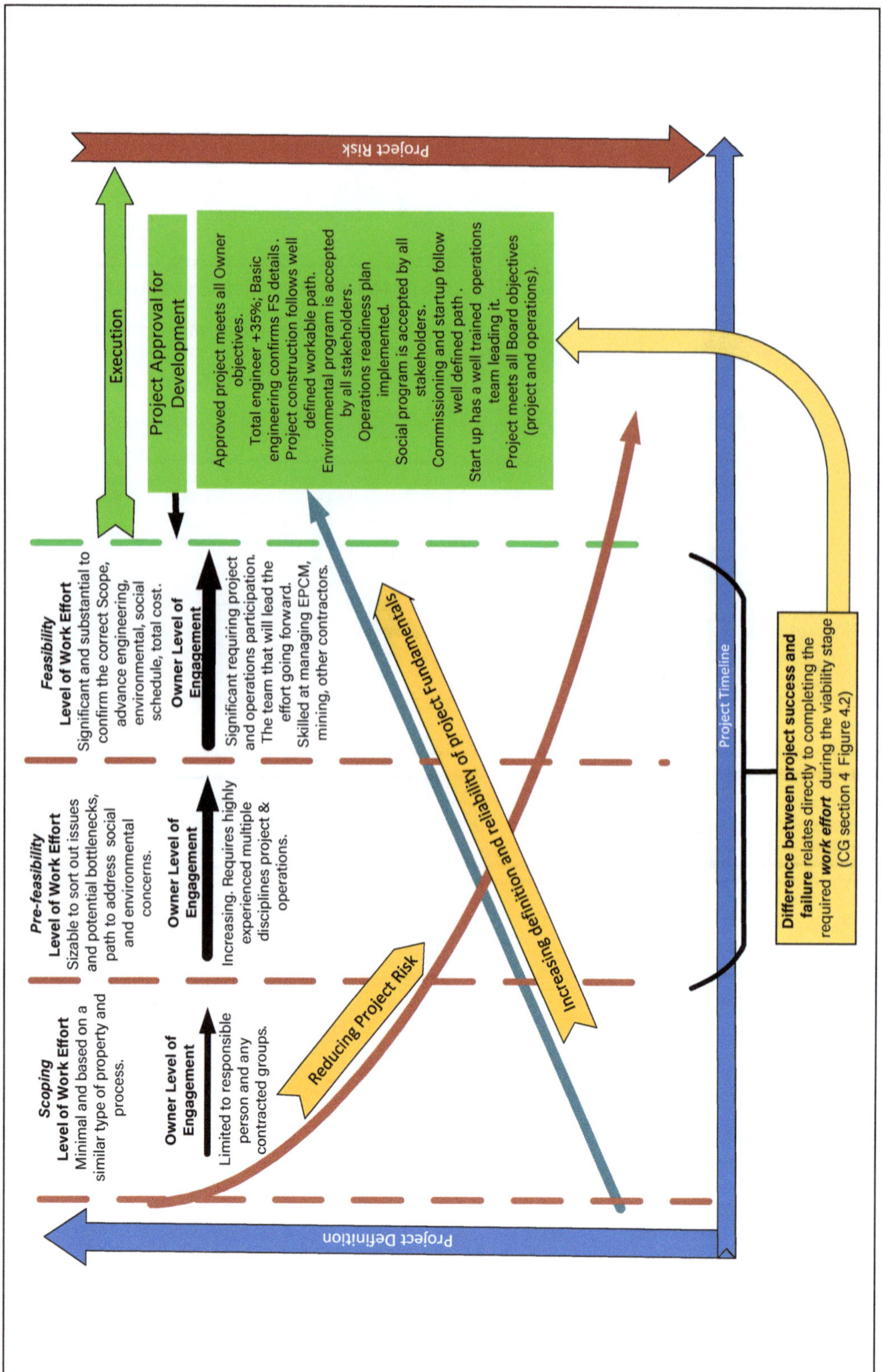

Figure 2.4 Maximize probability of project success

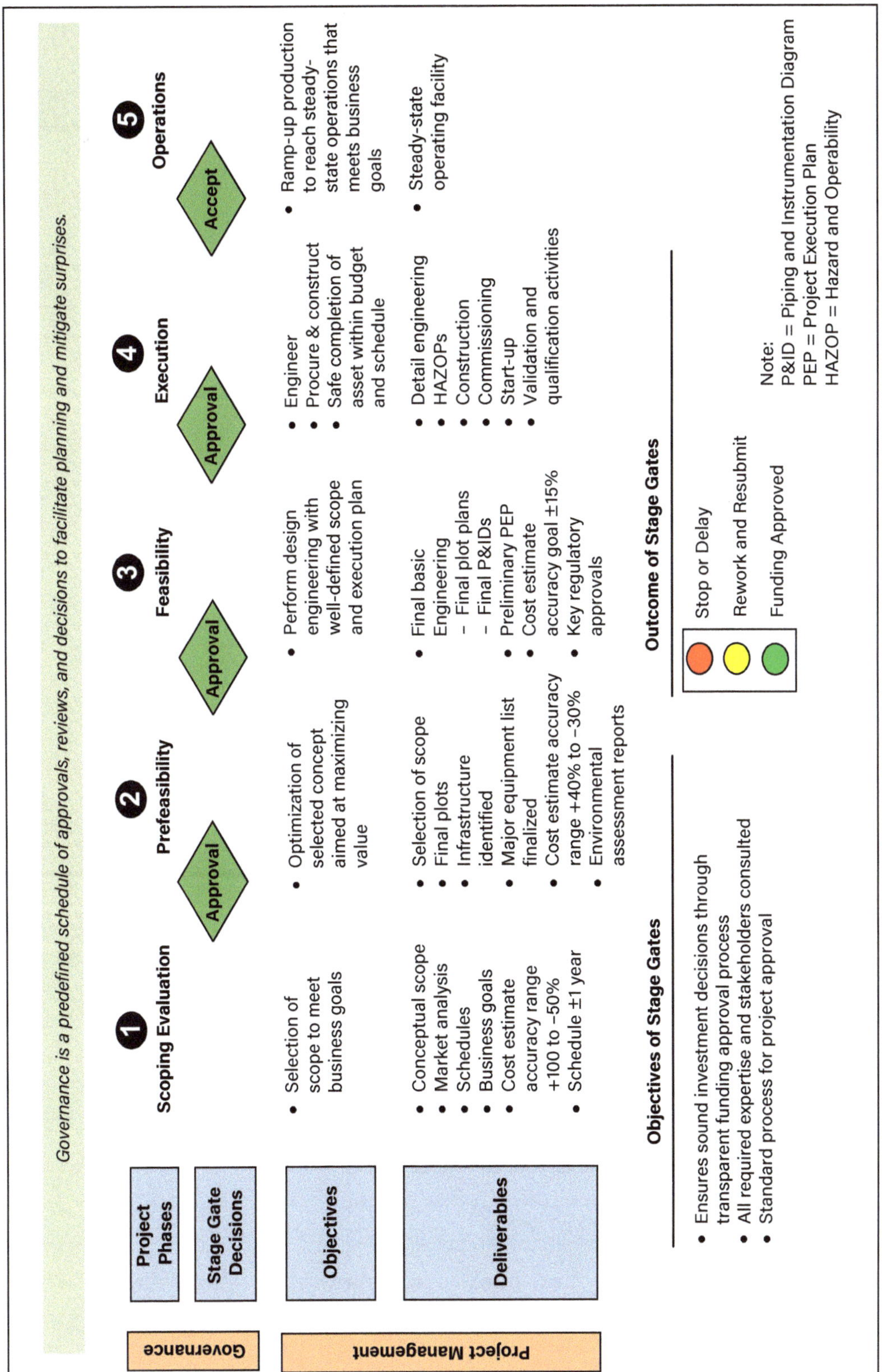

Figure 2.5: Governance program (stage gate)

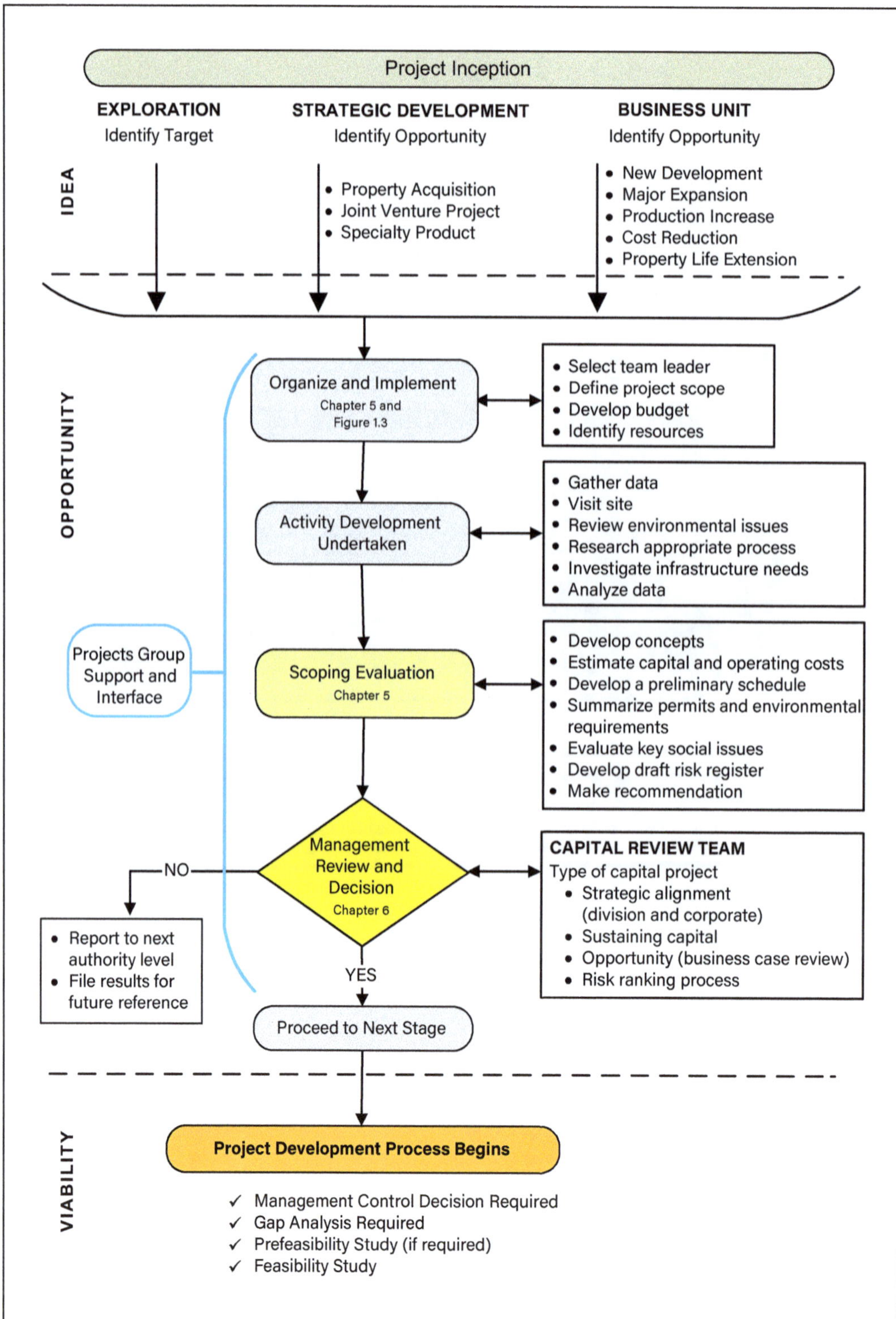

Figure 2.6: Opportunity activities

UTILIZING THE GUIDEBOOK

No.	Item	Status	Date
1	Understand definition of a project (*PMfM* Chapter 1; p.5)		
2	Understand stage and components of project as it stands		
3	Utilize *PMfM* Figure 1.1 to develop gap analysis before taking next steps		
4	Understand components of project development *PMfM* Flowchart 1.1		
5	Study referenced *PMfM* chapters to gain clarity for next steps		
6	Ethics apply *PMfM* Chapter 3		
7	Governance apply *PMfM* Chapter 4		
8	Management buy in ["stage gate"] *PMfM* Chapters 6,16,17		
9	Project process for studies (scoping through feasibility)		
	A. Background summary		
	B. Concise project description, including boundaries		
	C. Project phase objectives		
	D. Scope of work, going forward—expected deliverables		
	E. Scope of services required		
	F. Resources required		
	G. Stakeholders (internal and external) and partners		
	H. Timeline and budget		
	I. Approvals required		
10	Recommendation for project management of next stage—resolved		
11	Contracted services requirements defined and in place		
12	Authorization for Funds Expenditure (AFE) for Study completed and approved		
	• AFE guidelines review—undertaken		
	• AFE form—prepared and signed		

SECTION 3

Project Management Activity

OBJECTIVE

It is necessary first to identify the steps that both the Project Advocate and the assigned PM are responsible for undertaking when placed in charge of a project. Together, these steps focus the process on the elements requiring attention, regardless of the stage of the project.

It is critical that the Owner's team understands each of the complex components of mining projects and the best practices for managing their progression through the five stages of project life.

Each section of these guidelines is designed to direct the reader to the corresponding chapter in *Project Management for Mining* (*PMfM*), which must be thoroughly understood for successful implementation. *PMfM* must be used alongside these guidelines.

REFERENCES

Item	*PMfM*	Page
Project Management	Chapter 1	5
Project Control	Chapter 24	427
Project Charter	Chapter 38	700
Project Execution Plan	Chapter 18	277

CG Figure	Figure Caption	*PMfM* Reference
Figure 3.1	Overview of mining project stages and processes	New Figure
Figure 3.2	Project management activity	Figure 1.1, p.6
Figure 3.3	Study setup—implementation	New Figure

Checklist Title	*PMfM* Reference
Project management activity	New Checklist

SCOPE

Organizing

Projects rely on well-organized and fit-for-purpose information. Key areas of focus include the following:

Project Elements

Setting up a project requires the project manager to understand all the elements of the project and who the stakeholders are. Understanding each stakeholder's objectives, and the project's impacts—both positive and negative—is not always an easy or straightforward task.

Needs Assessment

- It is the Advocate's and PM's responsibility at this point to collect all the data and review it, no matter how small or whether they are told, "Do not worry about this."
- If this is a process change to the existing plant, then everything from the ore body to the tailings needs to be reviewed to understand the true and complete scope of the project.
- How will this project be managed, and who will be involved (Figure 3.1)?

Planning Process

- Create a work plan—a set of project documents—that provides the Advocate and PM with a method to identify the scope to be studied, identify the resources required, plan the activities to execute the scope, and support the funding request (likely an AFE, depending on the size).
- For the study phases ranging from Opportunity to Viability, this is a Project Charter. It has a scope and work plan to advance the study and can also be used on smaller brownfield projects (*Companion Guidebook (CG)* Section 18).
- It should be noted that in the 30 years since the first *PMfM* was published, a Project Charter document has become more prevalent within the mining fraternity. Further, today's Project Charter is no longer a bureaucratic duplicate within the Governance process (as described on page 35 of *PMfM*) but instead now functions usefully as a tool to be created ahead of the Feasibility Study, serving as an abbreviated precursor to the Project Execution Plan. See *CG* Section 7 for more Charter content details.
- For the Execution Stages, the Project Execution Plan (PEP) is created during the Feasibility Stage and is the work plan for large-scale projects (*CG* Section 12).
- Determine what the organization for the project will require and set up a hiring plan.
- Ensure that all controlling information (scope, schedule, budget, and contracting plan) is assembled.

Implementation

Once the correct information has been collected, evaluated, and organized, decisions need to be made on how to proceed at the various stages of the project.

Execute Work Plan Activity

The PM now has all the relevant documents and plans to move forward and begins the implementation of activities.

Establish Project Controls

Setting up the controls before the project is executed is important to staying ahead of the project as activity gets underway.

Performance Management

It is critical to ensure that all the contractors and Owner groups on the project see themselves as "One Team with One Goal." Once aligned, they can then manage the issues.

Overview of Mining Project Stages and Key Activities

It is critical that the senior project advocate in the company and the PM maintain a holistic view of the project as it advances. Figure 3.1 depicts how the multiple stages and key activities intersect as the project advances from one stage to the next.

- It is important to keep both operations and support groups involved at the appropriate times.
- Review sessions need to be held on technical developments and sustainability (Environmental, Social, Regulatory [*CG* Section 10]).
- Ensure the safety of all personnel is maintained during all work in the field for all stages of the project.
- Develop financial requirements as the project moves forward.
- Conduct management reviews and approvals to move forward, pause for more field activities, or shut it down as "not viable."

Project Management References

Additional support for those new to managing projects:

- Project Management Institute (PMI)—This will help in learning the fundamentals of project management.
- Association for the Advancement of Cost Engineering (AACE) [cost engineering and project controls professionals]
- Construction Industry Institute (CII)

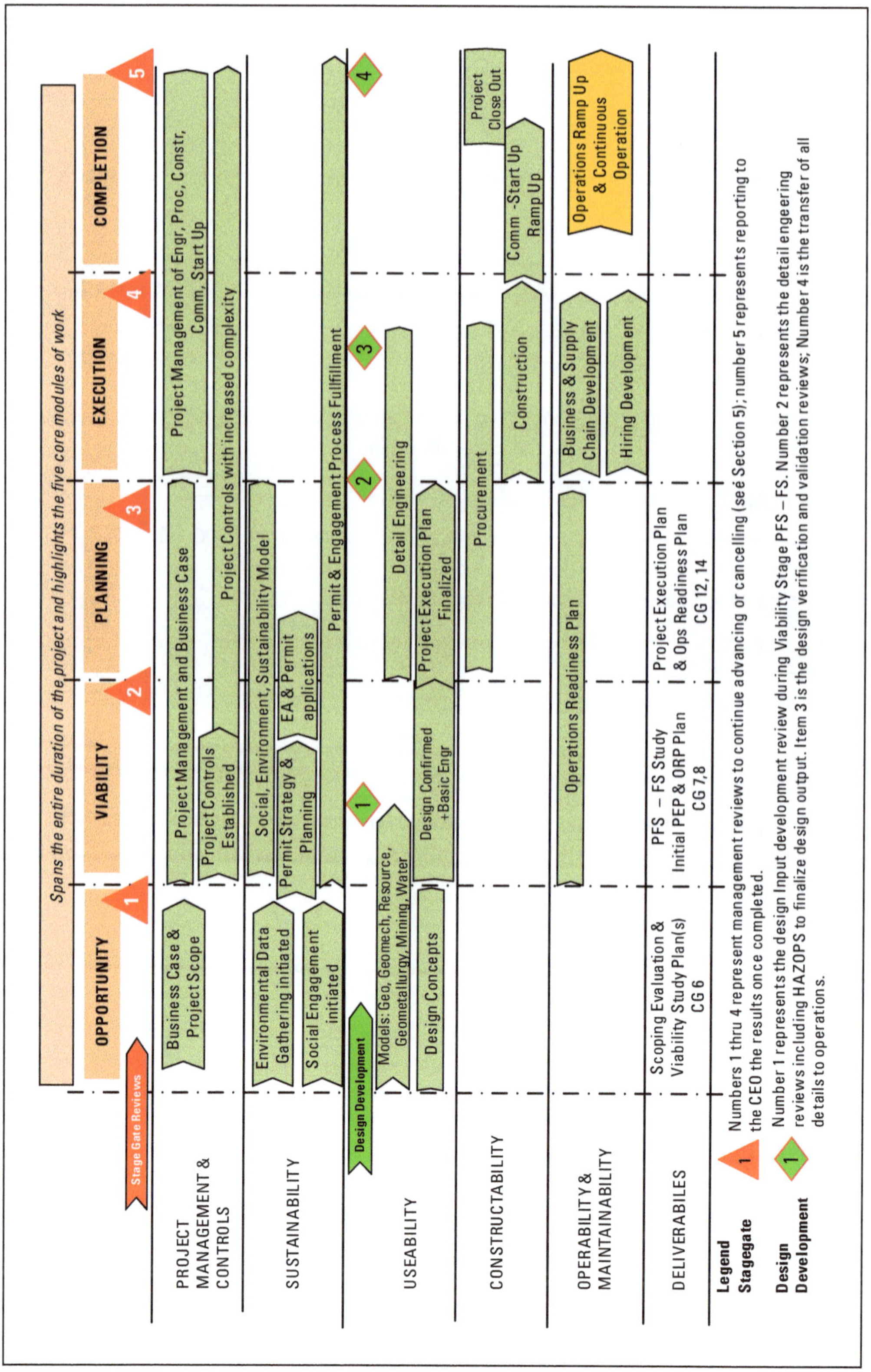

Figure 3.1: Overview of mining project stages and processes

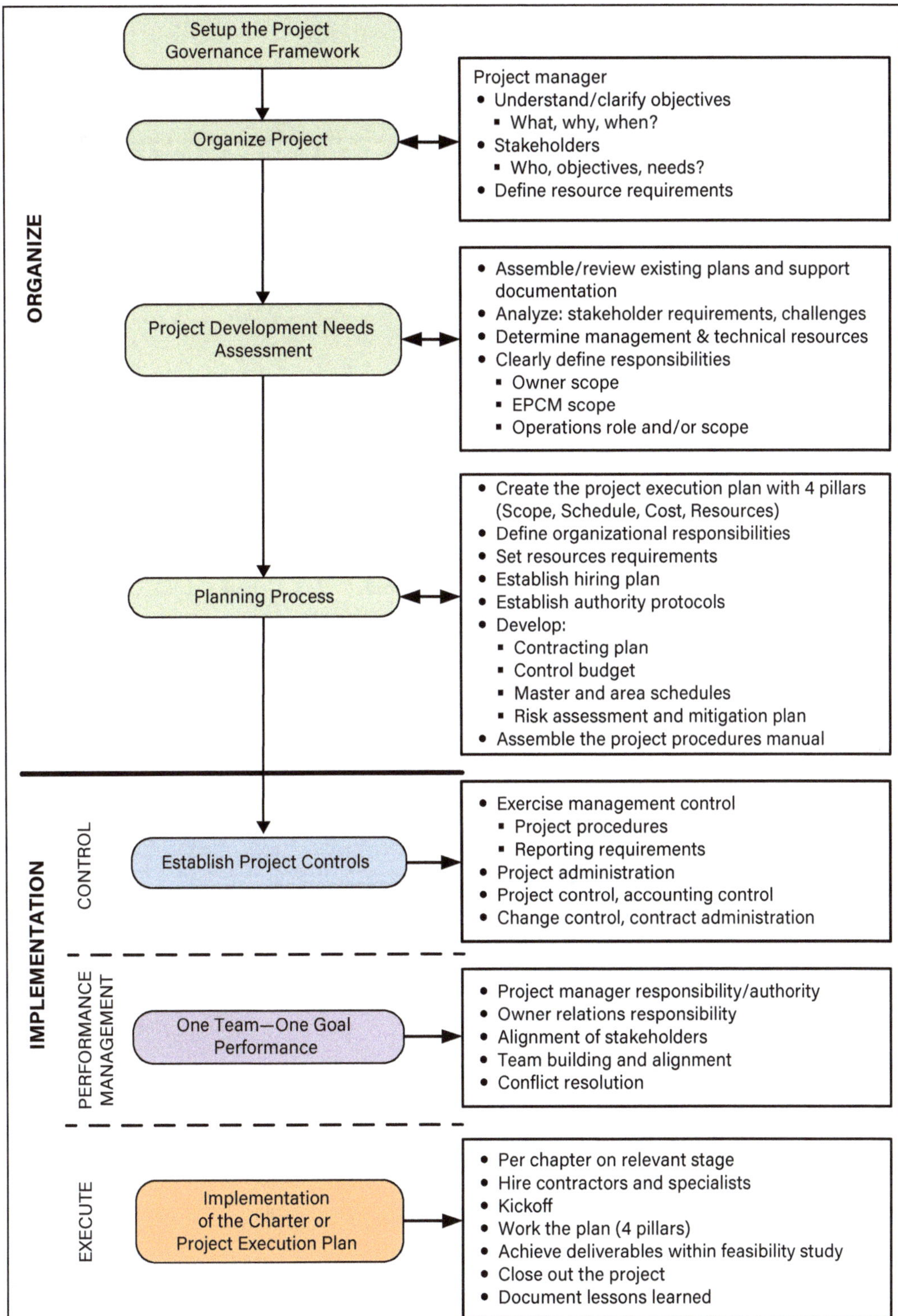

Figure 3.2: Project management activity

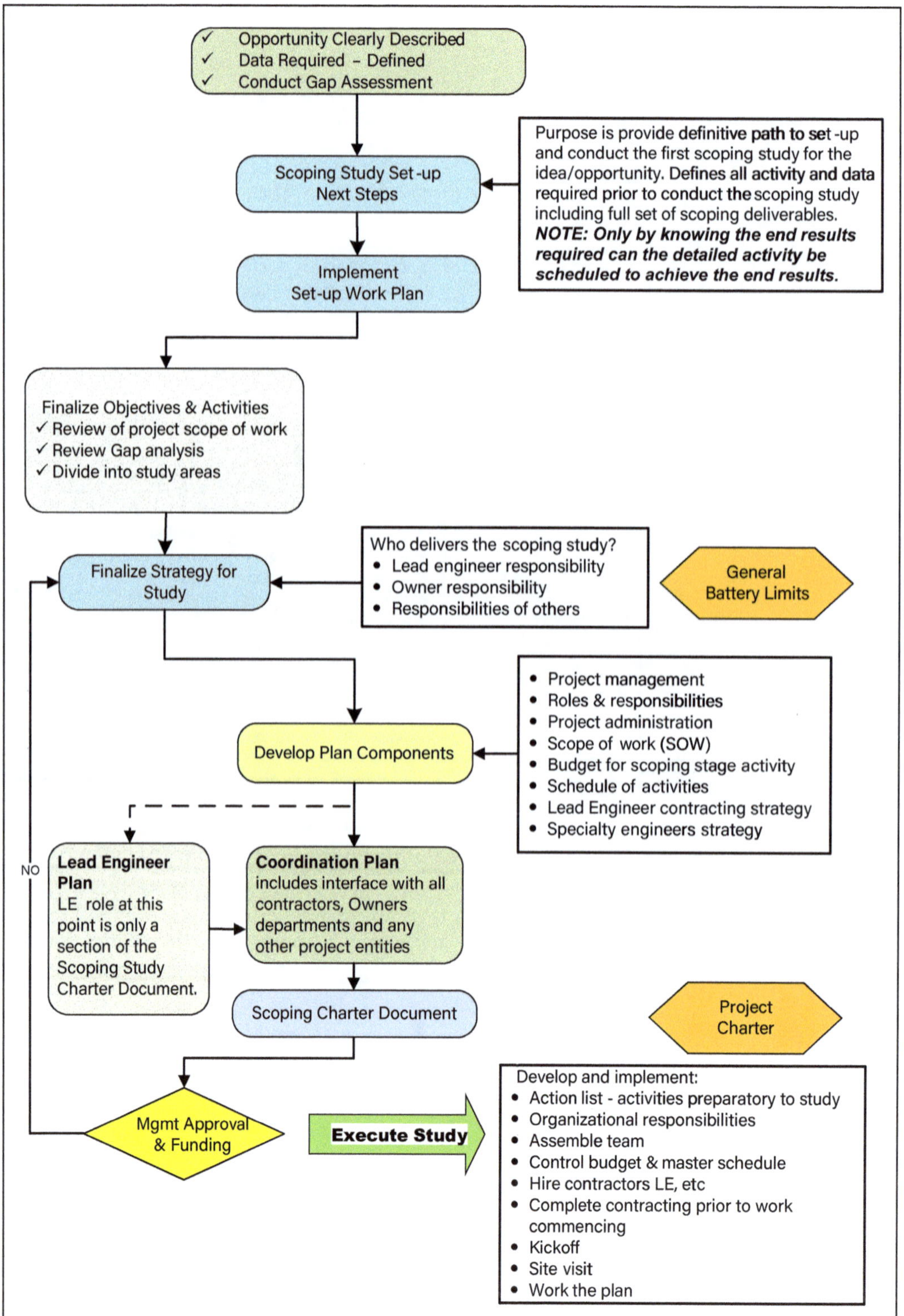

Figure 3.3: Study setup—implementation

PROJECT MANAGEMENT ACTIVITY

No	Item	Status	Date
1	Understand definition of a project (*PMfM* Chapter 1)		
2	Utilizing simple project charter framework to describe the project scope, boundaries, etc as it is understood.		
4	Review for clarity components of project development *PMfM* Flowchart 1.1, p.18		
5	Project advocate and project manager immediately develop simple project charter for studies		
	A. Background summary		
	B. Concise project description, including boundaries		
	C. Project phase objectives		
	D. Scope of work, going forward—expected deliverables		
	E. Scope of services required		
	F. Resources required		
	G. Stakeholders (internal & external) and partners		
	H. Timeline and budget		
	I. Approvals required		
6	Utilize Figure 1.1 to develop gap analysis before taking next steps		
7	Study referenced *PMfM* Chapter 1 to gain clarity for next steps		
8	Ethics *PMfM* Chapter 3		
9	Governance *PMfM* Chapter 4		
10	Update project charter for next step study		
11	Management buy in ["stage gate"] *PMfM* Chapter 6,16,17		
12	Recommendation for project management of next stage—resolved		
13	Contracted services requirements defined and in place		
14	AFE for study completed		
	AFE guidelines review—undertaken		
	AFE form—prepared and signed		

SECTION 4

Level of Effort Required per Project Stage

OBJECTIVE

When advancing through the project stages, it is important to begin each stage of development with a clear understanding of the level of work required to complete that stage. This is often referred to as "starting with the end in mind," and it is therefore useful to understand all the elements that must be included in that process.

Each section of these guidelines is designed to direct the reader to the corresponding chapter in *Project Management for Mining (PMfM)*, which must be thoroughly understood for successful implementation. *PMfM* must be used alongside these guidelines.

REFERENCES

Item	*PMfM*	Page
Project Stage Work Efforts	Chapter 15	243

CG Figure	Figure Caption	*PMfM* Reference
Figure 4.1	Summary comparison of study work efforts	Figure 15.1, p.244
Figure 4.2	Comparison of project stage work efforts	Figure 15.2, p.245
Not Included	Comparison of PMfM stages versus FEL/FEED stages	Figure 15.3, p.264

SCOPE

Work Effort Requirements

Figure 4.1 provides a summary comparison of the escalating levels of work effort required at the formal study stages of project life. These levels of effort are the minimums required to properly prepare the project for successful execution and to inform correct investment decisions.

Figure 4.2 is the 17-page *PMfM* Figure 15.2, providing a more in-depth, multipage comparison. If the activities are undertaken to the degree listed, then gaining approval to move on to the next project stage will be streamlined.

This *Companion Guidebook (CG)*, as well as *PMfM*, adheres to the nominal standards of studies, including Scoping Evaluation, Prefeasibility Study (PFS), and Feasibility Study (FS). It is acknowledged that Engineering, Procurement, and Construction Management (EPCM) firms also use FEL (front-end loading) as a process. Refer to *PMfM* Chapter 15, p.262 to understand how FEL fits with the nominal standards used in *PMfM*.

For each of the study phases—Scoping, Prefeasibility, and Feasibility—there is a level of effort that represents the common industry minimum work effort. Figure 4.1 provides the reader with a quick one-page summary of the activities required in each study stage. *PMfM* also includes a Definitive Estimate work effort that follows the FS and is where Basic Engineering is completed and Detailed Engineering is increased above 70%. See *PMfM* Figure 30.5, p.566 and *CG* Section 17.

Each study should include internationally recognized elements that outline the project. The following items are recommended:

- Purpose of Study
- Project Management
- Site Visits
- Study Document
- Location / Property Position
- Current Status and History
- Exploration / Geology
- Exploration Drilling
- Mineral Resource
- Reserves
- Mine Plan
- Metallurgy
- Engineering & Process Design
- Site Infrastructure
- Hydrology
- Communications
- Power
- Procurement
- Work Force
- Environmental
- Permits
- Social and Sustainability
- Risk
- Schedule
- Project Execution Status
- Capital Cost Estimate
- Operating Cost Estimate
- Market Plan and Market Potential
- Business Considerations
- Financing

MITIGATING RISK (Best Practices Approach for the Study Advancement Path)

Scoping Evaluation	Prefeasibility Study	Feasibility Study
Drilling Sufficient for a resource; mostly RC.	**Drilling** Initial infill of wide spaced drilling; includes core.	**Drilling** Holes spaced on a grid that satisfies CRIRSCO.
Resource Mostly inferred. Broad physical limits drawn.	**Resource** Indicated mostly, some measured; lithology map. Selected mineral samples; initial geologic model.	**Resource** 80% measured and indicated. Full geologic model. Detailed mineralogical sampling study performed.
Reserve No reserves.	**Reserve** Preliminary tons and grade meet NI 43-101, SEC S-K 1300 quality requirements.	**Reserve** Mine plan 80% proven for NI 43-101, SEC S-K 1300.
Mining Generic mining method assigned. Little or no data to support.	**Mining** Mining method selection based on geologic data. Rudimentary geotechnical and hydrology data.	**Mining** Detailed mining methods and mine plans configured. Site-specific geotechnical and hydrogeological data.
Metallurgy Minimal testing; recoveries estimated.	**Metallurgy** Core sampling; recoveries based on bench tests.	**Metallurgy** Test suite and pilot plant confirm recoveries and process.
Engineering & Process Design Preliminary, based on similar plants. Throughput capacity estimated. Infrastructure allowance. Recommended total engineering ~1%.	**Engineering & Process Design** Conceptual design; initial plot plans; simple GAs. Probable flowsheet; preliminary production rates. Trade-offs conducted. Infrastructure defined. 3-D CADD model initiated. Total engineering +7%.	**Engineering & Process Design** Drawings/specs complete for full definition of scope. Layout established; process frozen; production set. Trade-offs complete. Infrastructure finalized. Basic engineering +85%; total engineering +30%.
Environment/Permits/Social Review existing data. Environmental, permit and social concerns investigated.	**Environment/Permits/Social** Gather baseline data. List permits. Assess social. Environmental Management Plan (EMP) created.	**Environment/Permits/Social** Baseline data complete. Permits prepped. Social strategy complete. EMP implementation underway.
Field Construction No knowledge of construction issues.	**Field Construction** Some geotechnical drilling undertaken.	**Field Construction** Geotechnical sufficiently complete for construction.
Contingency Generic average +40%; range 30%–70%.	**Contingency** Broad evaluation delivers +22%; range 15% to 30%.	**Contingency** Detail undertaken yields +15%; range 12% to 18%.
Accuracy at 90% Confidence At best ±40%; range +100% to –50%.	**Accuracy at 90% Confidence** At best ±25%; range is from +40% to –30%.	**Accuracy at 90% Confidence** Goal should be ±15%; but range is +25% to –20%.

Figure 4.1 Summary comparison of study work efforts

SCOPING EVALUATION	PREFEASIBILITY STUDY	FEASIBILITY STUDY
PURPOSE		
• Preliminary investigation to support the pursuit or rejection of an identified opportunity. Potential commercial viability assessed. Estimate accuracy is suitable for project rejection, but not adequate for management go-ahead of construction.	• Study of alternative project configurations to demonstrate the technical and economic feasibility of the preferred conceptual installation (type and size) that could realistically be built. Yields the framework for a follow-up feasibility study.	• Describes the complete, optimized project installations to be built. Sufficient scope definition for Board approval, financing, budgeting, and control. Estimate accuracy suitable for nonrecourse external financing (i.e., a bankable document).
PROJECT MANAGEMENT		
• Mostly internal Owner personnel.	• Mix of engineers and specialists from an outsource EPCM and the Owner.	• Team mostly from EPCM, but support from external specialists and Owner.
• Project leader—a person familiar with this minerals industry sector and experienced in project management.	• Project team has relevant minerals industry knowledge. Leader familiar with Owner's objectives.	• Leader skilled in engineering management. Project team includes design engineers and estimators.
SITE VISITS		
• Single site visit by Owner's staff. • General gathering of germane data.	• Two plus visits by Owner specialists. • Preliminary review of host country business climate and social situation.	• Multiple on-site assessments. • Focused depiction of host country and community attitudes on the project.
STUDY DOCUMENT		
• Document produced by Owner staff or a single-sourced outside consultant.	• Document typically produced by an external engineering or EPCM firm.	• Study usually produced by EPCM firm. Sufficient definition for project control.
• Labor hours: 500 to 10,000. • Duration 4 to 15 months (average 8).	• Labor hours: 5,000 to 120,000. • Duration 8 to 36 months (average 18).	• Labor hours: 20,000 to 400,000. • Duration 12 to 39 months (average 27).
• Total cost: \$50,000 to \$2 million.	• Total cost: \$400,000 to \$20 million.	• Total cost: \$1.5 million to \$80 million. • Typically 4% ± 2% of total project cost.
LOCATION/PROPERTY POSITION		
• Site location map. • Regional climate data. • Property Ownership/royalty status: – Property lease desirable.	• Site map with general topography. • Site-specific climate data. • Property Ownership/royalty status: – Property under lease. – Claims list. – Mineral rights secured. – Encumbrances identified.	• Site map with detailed topography. • Site climate data over multiple years. • Property Ownership/royalty status: – Property title rights fully controlled. – Claims list and claim map. – Mineral rights secured. – Encumbrances identified.

Figure 4.2 Comparison of project stage work efforts (continues)

SCOPING EVALUATION	PREFEASIBILITY STUDY	FEASIBILITY STUDY
CURRENT STATUS AND HISTORY		
• Possibly none. Often just a new idea.	• Historical chronology. • Past production, if any.	• Historical chronology. • Past output, old workings maps, if any.
EXPLORATION/GEOLOGY		
• Data collection: From library review and a preliminary site visit.	• Data collection: From library review and by on-site geologist(s).	• Data collection: All on-site by geology team (Owner and contract).
• Review existing maps. Preliminary geologic mapping.	• Geologic maps and associated cross sections produced.	• Geologic maps, sections, level plans. Deposit defined in three dimensions. • If an underground project, access (e.g., shaft, adit, drift) is requisite.
• Outcrop samples taken. • Geophysics optional.	• Geophysical/geotechnical sampling. • Test pits.	• Geophysical/geotechnical sampling. • Test pits, trench, and channel samples.
• Preliminary geologic assessment.	• Basic geological assessment.	• Detailed geologic assessment.
• Limited assay data available.	• Preliminary mineralogical sampling.	• Detailed mineralogical mapping.
• Preliminary mineralogical review.	• Preliminary mineralogical study.	• Detailed mineralogical support study.
• Preliminary coal quality analysis.	• Preliminary coal quality/washability.	• Detail coal quality/washability study.
EXPLORATION DRILLING		
Drill hole parameters: • A few targeted holes (optional). • Mostly RC holes; for initial resource	**Drill hole parameters:** • Initial fill-in of wide-spaced holes. • Sufficient holes, including core holes for an NI 43-101 PEA.	**Drill hole parameters:** • Closely spaced holes, on a grid. • Sufficient holes for an NI 43-101, JORC, or SEC S-K 1300 reserve and model.
Checks: • Check on existing drill-hole data. • Drill-hole location map. • Spot confirmation assays (optional).	**Checks:** • Verification of drill-hole data. • Drill-hole location map. • Check assays of existing core. • Twin holes (optional). • Select confirmation mineral assays.	**Checks:** • Verification of drill-hole data and SG. • Collar and downhole surveys. • Full re-assay of existing core. • Twin holes (preferably >5). • Confirmation assays (1 in 20).
MINERAL RESOURCE		
Geologic model: • Broad physical limits drawn. • Rudimentary statistics.	**Geologic model:** • Initial model with derived physical limits from known geology, backed by basic statistics and variograms. • Lithology/tonnage factors interpreted.	**Geologic model:** • Fully complete model with physical limits set by variogram statistics and with the geology superimposed. • Lithology/tonnage factors fixed.
Resource: • Inferred mostly; some indicated.	**Resource:** • Indicated mostly, some measured.	**Resource:** • 80% measured/indicated.

Figure 4.2 Comparison of project stage work efforts (continued)

SCOPING EVALUATION	PREFEASIBILITY STUDY	FEASIBILITY STUDY
RESERVES		
• Assumed calculation parameters.	• Derived calculation parameters.	• Test-derived calculation parameters.
• Preliminary (Possible and Probable).	• Probable and Proven tons and grade that satisfy an NI 43-101 "PEA" or SEC S-K 1300 "Initial Assessment."	• NI 43-101, JORC, or SEC S-K 1300 "Technical Report" quality (Proven and Probable) tons and grade.
• No reserve able to be classified.	• Mine plan consists mostly of probable reserve status material.	• Mine plan contains a minimum of 80% proven reserves.
• Manual calculation (typically cross section or polygonal), i.e., conceptual estimate by the Owner.	• Reserve delineated sufficiently for a reasonable estimate of tons and grade. Generally a geostatistical block model but can be cross-sectional.	• Two reserve calculations; one has to be geostatistical, using a detailed block model. One must be by an external, recognized specialist.
MINE PLAN		
Mining method: • Generic surface or underground extraction method assigned.	**Mining method:** • Specific method established, based primarily on available geologic data.	**Mining method and mine plan:** • Detailed method and mine plan configured.
Mining parameters: • Minimal engineering basis. • Little or no data to support.	**Mining parameters:** • Rudimentary geotechnical data. • Rudimentary hydrology data.	**Mining parameters:** • Site-specific geotechnical data. • Site-specific hydrology data.
Dilution: • Ignored.	**Dilution:** • Generic factors assigned.	**Dilution:** • Specific for chosen mining method.
Pit slopes: • Generic 40° to 45°.	**Pit slopes:** • Preliminary by rock type.	**Pit slopes:** • By cell mapping and oriented core.
Economic parameters for mine plan: • Order of magnitude; ±200%.	**Economic parameters** for cones: • Preliminary; ±40%.	**Economic parameters** for cones: • Improved estimate; ±20%.
Process recovery for mine plan: • Typically assumed; ±30% accuracy.	**Process recovery** for floating cones: • Preliminary; ±10% to 15% accuracy.	**Process recovery** for floating cones: • Improved estimate; ±5% accuracy.
Cutoff grade: • Assumed.	**Cutoff grade:** • From floating cone parameters.	**Cutoff grade:** • Optimized using COG equations.
Surface pit design: • 40° to 45° generic sketch. • Simple outline of final pit.	**Surface pit design:** • Floating cones with pushbacks. • Haul roads factored into design. • Incremental and final pit outlines.	**Surface pit design:** • Detail optimized phase designs. • Haul road progressions in the design. • Phase, annual and final pit outlines.
Underground design: • General outline of development. • Metal: Assumed stoping system. • Coal: Assumed longwall or room and pillar; assign a percentage extraction.	**Underground design:** • Simple outline of development. • Metal: Specific stoping system. • Coal: Basic design for ventilation, roof control, and secondary mining.	**Underground design:** • Detailed outline of development. • Metal: Specific stoping system. • Coal: Detailed ventilation, haulage, roof control, utilities, and mine designs.

Figure 4.2 Comparison of project stage work efforts (continued)

SCOPING EVALUATION	PREFEASIBILITY STUDY	FEASIBILITY STUDY
Mine production schedule: • Paper schedule. • Mining rate from paper schedule.	**Mine production schedule:** • From pushbacks. • Mining rate from mine plan.	**Mine production schedule:** • Detailed annual schedules. • Mining rate from mine plan. • Optimized COG strategy.
• "Mine life" production tonnages (ore and waste) and "mine life" grade.	• Yearly and mine life tonnages, grade, production, and strip ratios.	• Yearly and mine life tonnages, grade, production, and strip ratios.
Waste dumps/coal refuse storage: • Potential area blocked out. • Simple outline of final shapes.	**Waste dumps/coal refuse storage:** • Design to match total waste tons. • Incremental and final outlines.	**Waste dumps/coal refuse storage:** • Sites and heights selected. • Phase, annual, and final outlines.
Mine services: Assumed.	**Mine services:** From sketch designs.	**Mine services**: Fully outlined.
Resource requirements: • Assumed	**Resource requirements:** • Generic equipment/consumables. • Personnel list.	**Resource requirements:** • Detailed equipment/consumables. • Personnel list and organization chart.
METALLURGY		
Ore sampling: • Generally none. Assumed values.	**Ore sampling:** • Metallurgical sampling of core.	**Ore sampling:** • Metallurgical sampling of core and of development workings.
Metallurgical mapping: • None.	**Metallurgical mapping**: • Sampling of ore body zones.	**Metallurgical mapping**: • Complete mapping of ore body.
Tests: Minimal or none. Mostly just a review and verification of existing metallurgical information.	**Tests:** Advanced bench-scale testing to determine preliminary recoveries.	**Tests:** Complete comprehensive suite of tests and pilot plant confirm design and derive defendable recoveries.
Recoveries: Estimated from already existing data or assumed benchmark.	**Recoveries**: Preliminary, based on actual metallurgical tests performed.	**Recoveries**: Calculated from the suite of metallurgical tests performed.
Technology selection: • Use industry typical.	**Technology selection**: • Compare project specific options.	**Technology selection**: • Identify optimum option.
Coal washability: Sink/float data: • Estimate.	**Coal washability**: Sink/float tests: • Preliminary.	**Coal washability**: Sink/float tests: • Improved estimate.
Plant production schedule: • "Mine life" ore tons, grade, recovery, and concentrate/metal output.	**Plant production schedule:** • Yearly ore tons, grade, recoveries, and concentrate/metal output.	**Plant production schedule:** • Yearly ore tons, grade, recoveries, and concentrate/metal output. • Ramp-up to full design by month.

Figure 4.2 Comparison of project stage work efforts (continued)

SCOPING EVALUATION	PREFEASIBILITY STUDY	FEASIBILITY STUDY
ENGINEERING AND PROCESS DESIGN		
Design basis: Preliminary; concepts extrapolated from similar plants. • First cut assumes estimated plant product(s) and throughput capacity. • Likely operating procedure postulated. • Preliminary model created (optional)	**Design basis**: General; key drawings for initial definition of scope. • Preliminary mine and process rates, plant product(s), and capacity. • Conceptual design engineering done. • Preliminary CADD 3-D models: civil, structural, piping, and mechanical. • Potential operating spaces selected.	**Design basis**: Complete; all design drawings for full definition of scope. • Final mine and process throughput rates, plant product(s), and capacity. • CADD 3-D models from all disciplines. • Basic engineering optimized and essentially complete; (>85%).
Value engineering: • None.	**Value engineering:** • Key trade-offs conducted and done. • Best case configuration selected.	**Value engineering:** • Design trade-offs all complete. • Final configuration is frozen.
Design concept: • Criteria and specifications outlined.	**Design criteria:** • Basic design criteria set. • Engineering codes selected.	**Design specifications:** • Process design criteria set. • General specifications complete.
Site plan: • Sketch of approximate locations.	**Site plan:** • Likely facility locations postulated. • Preliminary plot plans.	**Site plan:** • All facility locations established. • Detail plot plans.
Layout: • Approximate geographic locations with site location map. • Sketch illustrations.	**Layout:** • Some optimization of initial facility locations with site location map. • Simple GAs. • 2-D plant drawings. Initiate 3-D model.	**Layout:** • Optimal facility locations on site map. Topography and regulations respected. • Detail GAs for design development. • Conveyor profiles developed. • 3-D CADD model advanced.
Flow sheets: • Initial process design; process parameters set from industry typical. • Potentially viable flow sheet idea derived from known processes. • Generic assumptions of values. • Simple block flow diagram. • One process flow diagram (PFD).	**Flow sheets:** • Basic process parameters set from metallurgical testwork. • Probable flow sheet developed, based on preliminary testwork data, typically using a METSIM model. • Main elements of water, heat, and material balances. • Major equipment list only. • Major PFDs created.	**Flow sheets:** • Full suite of process parameters from metallurgical testwork. • Detailed flow sheet advanced, based on comprehensive pilot testwork using a model such as from METSIM. • Optimize water, heat, and mass balances. • Full, optimized equipment list. • All PFDs optimized and complete.
Engineering flow diagram: • None.	**Engineering flow diagram:** • Materials handling system selected. • Major instruments and valves.	**Engineering flow diagram:** • Materials handling system complete. • Optimized line sizes, instruments, and valves.

Figure 4.2 Comparison of project stage work efforts (continued)

SCOPING EVALUATION	PREFEASIBILITY STUDY	FEASIBILITY STUDY
HAZOP: • None.	**HAZOP:** • Optional fatal flaw analysis.	**HAZOP:** • First HAZOP should be undertaken.
Reagent and consumables usage: • Industry typical.	**Reagent and consumables usage:** • Preliminary.	**Reagent and consumables usage:** • Complete and optimized.
Surge capacity and stockpiles: • None.	**Surge capacity and stockpiles:** • Preliminary estimates.	**Surge capacity and stockpiles:** • Detail calculations. • Ore-blending requirements addressed.
Operating control philosophy: • None.	**Operating control philosophy:** • Outline.	**Operating control philosophy:** • Detailed for all systems.
Resource requirements: • Assumed.	**Resource requirements:** • Initial equipment. • Personnel lists.	**Resource requirements:** • Detailed equipment and consumables. • Personnel and transport requirements.
Civil work: • Rough site topographical maps. • Special features noted • Possible site layout sketch (optional). • Access road, pit, TSF, dumps shown.	**Civil work:** • Topographical maps showing existing features (e.g., 1:10,000 with 10 m contours). • Rough geotechnical and soils reports. • Basic preliminary quantities. • Preliminary site and layout drawings. • Preliminary civil model used for GAs. • Laydown yards/building envelopes.	**Civil work:** • Semi-detailed topographical maps (e.g., 1:2,000 with 2 m contour intervals). • Semi-complete geotechnical report on collected site ground strength data. • Detail loadings and quantities. • Defined site, mine, and TSF plot plans. • Roadways, benches, ditches, culverts, channels, retention basins defined.
Equipment specifications: • Major equipment list. • General concept outline. • No specifications or data sheets.	**Equipment specifications:** • Major equipment list, including brief description of main items. • Initial sizings and types (optional). • Outline of criteria. • Preliminary specifications.	**Equipment specifications:** • Complete equipment list with all major equipment priced FOB job site. • Detail sizing. • Detailed specifications. • Data sheets on major equipment.
Architectural: • Generally none; sometimes sketches.	**Architectural:** • Sketches. • Relevant building codes identified.	**Architectural:** • Major architectural designs only. • Exterior elevations set.
Structural: • Site design conditions.	**Structural:** • Minimal, outlines only. • Basic structural steel and concrete layouts.	**Structural:** • Preliminary, some detailed. • Preliminary concrete foundations. • Buildings laid out showing platforms, stairs, and containment areas.

Figure 4.2 Comparison of project stage work efforts (continued)

SCOPING EVALUATION	PREFEASIBILITY STUDY	FEASIBILITY STUDY
Mechanical: • Preliminary major equipment identified.	**Mechanical:** • Major equipment types/sizes listed. • Equipment model initiated. • Preliminary locations for equipment.	**Mechanical:** • Full equipment list with specifications. • Equipment model well underway. • Recommended bidders list. • Key vendor information solicited.
Piping/P&ID/HVAC: • None.	**Piping/P&ID/HVAC:** • Single-line piping drawings; major pipeline sizes on key preliminary P&IDs. • Preliminary list of all large-diameter piping requirements. • Preliminary process piping model.	**Piping/P&ID/HVAC:** • Detail for major systems; general for rest. Schematics and P&IDs complete. • Detail list of all process pipeline sizing requirements (>2 inch diameter). • Valve and materials specifications. • Piping model well underway. • Initial HVAC, fire protection, plumbing.
Tankage: • None.	**Tankage:** • Preliminary list with tank sizes.	**Tankage:** • Complete tank list and data sheets.
Electrical distribution: • None.	**Electrical distribution:** • Basic one-line diagram. • Electrical rooms defined. • Preliminary equipment list.	**Electrical distribution:** • Design one-line diagrams complete. • Electrical control rooms designed. • Detailed equipment list. • Initial electrical equipment model.
Motors: • None.	**Motors:** • General description. • Preliminary sizing.	**Motors:** • Detail list of major items with hp. • Detailed sizing.
Instrumentation: • None.	**Instrumentation:** • General.	**Instrumentation:** • Preliminary, with supporting list.
Information systems: • None.	**Information systems:** • Preliminary.	**Information systems:** • Details on key systems.
Process description: • General.	**Process description:** • Narrative.	**Process description:** • Detailed.
Level of Definition: i.e., Percent of engineering completed, expressed as a percent of total engineering. • Recommend 1%; range <1% to 2%.	**Level of Definition**: i.e., Percent of engineering completed, expressed as a percent of total engineering. • Recommend >7%; range 5% to 12%.	**Level of Definition**: i.e., Percent of engineering completed, expressed as a percent of total engineering. • Recommend >30%; range 25% to 40%.

Figure 4.2 Comparison of project stage work efforts (continued)

SCOPING EVALUATION	PREFEASIBILITY STUDY	FEASIBILITY STUDY
SITE INFRASTRUCTURE		
• General overview provided.	• Likely facility locations postulated. • All critical facilities defined. • Major utility tie-ins identified.	• All facility locations finalized. • All support facilities listed. • Major utility tie-ins located.
HYDROLOGY		
• Regional water source(s) discussed. • No hydrology study. • Plant water balance not addressed. • Mine dewatering issues identified.	• Potential water source(s) identified. • Preliminary hydrology study. • Preliminary plant water balance. • Mine dewatering needs estimated.	• Water availability confirmed by study. • Local groundwater regime defined. • Detailed plant water balance. • Dewatering strategy confirmed.
COMMUNICATIONS		
• None.	• Communication systems study undertaken. • Satellite usage investigated.	• Communications licensing and standards issues built into design. • Designed for Owner WAN.
POWER		
• Industry-typical data utilized. • Overview of power availability. • Published regional power costs.	• Preliminary requirements derived. • Power sources/locations identified. • Distribution to site addressed (e.g., by using an ETAP model). • Electrical loads estimated. • Unit costs from local area.	• Power requirements confirmed. • Specific power source identified. • Pole line routing to site engineered. • Electrical loads and demand derived. • Unit costs from power company bid.
PROCUREMENT		
Logistics: • Review of existing logistics systems that could support the project. • Preliminary report highlighting significant obstacles to materials flow.	**Logistics:** • Initial logistics study made specifically for construction and operations. • Highlight report of material flow impediments to and from site. • Identify customs issues.	**Logistics:** • Detailed logistics plans produced. • Shipping, transportation plan, and marshalling requirements established. • Customs issues ascertained.
Materials procurement: • None.	**Materials procurement:** • Identification of appropriate parties to undertake materials purchasing and expediting for the construction effort. • Likely major equipment vendors listed.	**Materials procurement:** • Material and equipment sourcing plan outlined. Plans identify the single point entity responsible for sourcing. • RFQs to major equipment vendors. • Long-lead-time items identified.
Warehousing: • None.	**Warehousing:** • Location and size of major warehouse facilities estimated.	**Warehousing:** • Facilities for construction and operations identified.
Purchasing commitment: • Zero.	**Purchasing commitment:** • Zero.	**Purchasing commitment:** • Key long-lead-time equipment only.

Figure 4.2 Comparison of project stage work efforts (continued)

SCOPING EVALUATION	PREFEASIBILITY STUDY	FEASIBILITY STUDY
WORKFORCE		
Project management staffing: • Project Manager selected by Owner to champion the opportunity.	**Project management staffing:** • Project Management team defined: • Highly qualified Project Manager for project execution selected.	**Project management staffing:** • Owner's Project Management team roles listed; organization chart derived: • Proven successful Project Manager for project life span in place.
Project construction labor force: • Generic benchmark estimate.	**Project construction labor force:** • Preliminary estimate from research. • Data from known local constructors.	**Project construction labor force:** • Estimate derived from likely bidders. • Productivity based on regional survey.
Operations workforce levels: • Factored estimate, using area history. • Local availability of operations personnel reviewed.	**Operations workforce levels:** • Preliminary estimate. • Conceptual organization structure. • Operations personnel requirements listed by department. – Potential labor source identified. • Conceptual recruitment plan.	**Operations workforce levels:** • Zero-based labor force budget. • Detailed organization structure. • Detailed operations personnel list by job function and pay rate annually. – Source of labor identified. – Labor buildup schedule set. • Training for start-up described.
ENVIRONMENTAL		
Data review and collection: • Preliminary site visit is desirable. • Evaluate project for environmental and permit constraints. • Review of readily available data, e.g., country reports, library search. • Decision on what baseline studies to undertake.	**Data review and collection:** • Site visit(s) necessary. • Preliminary evaluation of impact on the environment. • Evaluate existing databases and country reports. • Long duration field studies begun. • Initiate baseline data collecting.	**Data review and collection:** • Multiple site visits necessary. • Focused characterization of potential environmental impacts. • All environmental databases and country reports reviewed. • Long-duration field studies underway. • Baseline data gathering complete.
Data assessment and preparation: • Identification of environmental "red flag" issues. • Overview of major environmental and permit issues. Characterize risks.	**Data assessment and preparation:** • Assess political/community issues. • Preliminary plans for handling environmental "red flag" issues. • Summary of the environmental and permit issues. Characterize risks. • Project design constraints identified.	**Data assessment and preparation:** • Assess political/community issues. • Full evaluation of environmental and permitting "red flag" issues. • Comprehensive environmental and permit report. Risk characterization. • Project design accommodates known environmental and permit constraints.

Figure 4.2 Comparison of project stage work efforts (continued)

SCOPING EVALUATION	PREFEASIBILITY STUDY	FEASIBILITY STUDY
Environmental management plan: • Conceptual plans for managing any identified "red flag" issues.	**Environmental management plan:** • Preliminary environmental management plan. (Equivalent scope of an ASTM "Phase I Assessment.") • Preliminary impact mitigation study. • Monitoring program prepared. • Geotechnical stability review. • Sediment and erosion control strategy. • Preliminary mine waste plan. • Initial tailings management plan.	**Environmental management plan:** • Comprehensive environmental management plan. (Equivalent scope of an ASTM "Phase II Assessment.") • Impact mitigation plan produced. • Monitoring program finalized. • Geotechnical stability analysis done. • Sediment and erosion control plan. • Mine waste management plan. • Detail tailings management plan.
• Minimal detail closure plan.	• Conceptual reclamation plan. • Evaluation of acid rock drainage potential from waste rock dumps. • Hazardous materials investigations.	• Reclamation plan. • Acid rock drainage potential study. • Hazardous materials plan.
	• Spill prevention strategy. • Draft ESIA/EIS/EA preparation initiated, as applicable.	• Spill prevention and response plan. • QA/QC parties responsible for environmental compliance identified. • Draft ESIA/EIS/EA well defined, and usually submitted to the regulators. • Serious public objections identified.
PERMITS		
Permit overview: • General overview of the environmental and other necessary permits gathered from readily available information.	**Permit overview:** • Comprehensive overview, understanding, and listing of the necessary permits for this mine type at this locality.	**Permit overview:** • Detailed evaluation of environmental and other permit requirements identified and highlighted within a "legal constraints" study.
Regulatory framework: • Framework investigated.	**Regulatory framework:** • Framework understood. • Permit acquisition timetable created. • Regulatory requirements communicated to the Project Manager and the EPCM deliverer company. • Ensure engineering design compatible with permit requirements.	**Regulatory framework:** • Framework understood. • Permit timetable in hand. • Selected governmental permit actions initiated; those possible from the level of engineering completed. • Permit acquisition support identified. • Key construction permits applied for.
Reporting: Requirements researched.	**Reporting**: Requirements understood.	**Reporting**: Obligations documented.

Figure 4.2 Comparison of project stage work efforts (continued)

SCOPING EVALUATION	PREFEASIBILITY STUDY	FEASIBILITY STUDY
SOCIAL AND SUSTAINABILITY		
Local community social issues: • Data compiled on- and off-site on existing significant social constraints. • Key stakeholders list assembled. • Potential risks characterization made.	**Local community social issues:** • Overview social impact assessment of the major risks and perceptions made. • Stakeholder outreach conducted. • Preliminary social management plan.	**Local community social issues:** • Comprehensive social impact assessment for project life produced. • Meetings with community and native cultures to assess trust or lack thereof. • Mitigation proposals to allay concerns.
Sustainability: • Data gathered on- and off-site on the locale's sustainability challenges. • Possible opportunities listed.	**Sustainability:** • Corporate and consultant review to ensure sustainability being addressed. • Local focus group interviews held.	**Sustainability:** • Study reviewed to ensure a net positive outcome for the ecosystem. • Responsibilities for outcomes listed.
RISK		
Risk characterization: • General overview of high-level risks, e.g., resource, locale, permits, social.	**Risk evaluation:** • Narrative on key risks; may include a preliminary fatal flaw analysis.	**Risk evaluation:** • Formal analysis using Monte Carlo simulation for totality of project risks.
Risk issues: • Gap analysis to show any disparity in current situation and corporate needs.	**Risk issues:** • Mitigation options to key risks identified and preliminarily costed. • List potential barriers to project. Optional informal fatal flaw analysis.	**Risk issues:** • Risk issues now fully understood and viable control measures generated. • Few unknowns remain, if any. Informal fatal flaw analysis.
SCHEDULE		
Mine development plan: • Development period assumed.	**Mine development plan:** • Development schedule estimated.	**Mine development plan:** • Schedule broken down in detail.
Project master schedule: • Level 1 Schedule. • Estimate outline only; one-page Gantt bar chart of major work elements.	**Project master schedule:** • Level 2 Schedule required. • Gantt chart showing overall time frames with key milestones. • Preliminary schedule calendar set. • Schedule outline for engineering.	**Project master schedule:** • Level 3 Schedule; logic-driven CPM required. • Gantt bar chart showing activity details and overall time frames. • Schedule calendar set with Owner. • Deliverables detailed in schedule. • Major milestones all identified.
Construction schedule: • Not required.	**Construction schedule:** • Optional, but desirable.	**Construction schedule:** • Semi-complete is necessary. • Critical path activities and schedule outlined and resourced.
Commissioning and ramp-up: • Not required.	**Commissioning and ramp-up:** • Outlined.	**Commissioning and ramp-up:** • Monthly critical path developed.

Figure 4.2 Comparison of project stage work efforts (continued)

SCOPING EVALUATION	PREFEASIBILITY STUDY	FEASIBILITY STUDY
PROJECT EXECUTION STATUS		
Project Execution Plan: • PEP not required. • Resources to conduct next project phase identified.	**Project Execution Plan:** • Conceptual PEP typically initiated, but not required. • Operations personnel viewpoints solicited. Buy-in obtained throughout. • Resources to conduct next project phase identified.	**Project Execution Plan:** • Preliminary PEP required. • Execution approach defined. • Operations buy-in of design and critical concepts firmly established. • Resources for project execution phases identified.
Project framework: • Project scope—broad description. • Monthly progress report published.	**Project framework:** • Project scope—preliminary outline. • Project design basis postulated. • Monthly progress report published. • QA/QC program outline.	**Project framework:** • Project scope—defined. • Project procedures manual complete. • Project design basis agreed to. • Change control procedure described. • Project controls system outlined. • Project reporting needs defined. • Monthly progress report published. • QA/QC program final. • Workplace safety program outlined. • Partnering philosophy statement. • Project audit philosophy established.
Construction contract strategy: • Likely unaware of construction issues. • Ground conditions mostly unknown.	**Construction contract strategy:** • Options outline discussion. • Geotechnical work initiated.	**Construction contract strategy:** • Decision as to whether lump sum, T&M , fee plus, GMP, etc. • Philosophy stated and justified. • Critical geotechnical work and test pits complete in all areas. Data in hand.
Percent constructed: • Zero.	**Percent constructed:** • Zero.	**Percent constructed:** • Zero, unless "fast track" project. (Note: Fast track not recommended).
CAPITAL COST ESTIMATE		
Basis: • Analogous, using mostly historical data from similar projects but with some expert judgment for gap areas.	**Basis:** • Parametric, using takeoffs and unit prices where data available, but much factoring and top down to fill in gaps.	**Basis:** • Bottom up mostly; virtually all task based by spread. Very little top down.
Sustaining CAPEX: • "Top down" order of magnitude solely. • Replacement capital items factored.	**Sustaining CAPEX:** • Mix of "bottom up" and "top down." • Replacement capital items factored.	**Sustaining CAPEX:** • Mostly "bottom up" for initial 5 years. • Replacement capital items calculated.

Figure 4.2 Comparison of project stage work efforts (continued)

SCOPING EVALUATION	PREFEASIBILITY STUDY	FEASIBILITY STUDY
WBS: Outlined. • Project code of accounts—none.	**WBS**: Preliminary. • Code of accounts—preliminary.	**WBS**: Defined. • Code of accounts—defined.
Mine capital cost estimate basis: • All factored (or expert judgment).	**Mine capital cost** estimate basis: • Partly from preliminary equipment list; rest from manuals and in-house data.	**Mine capital cost** estimate basis: • Detailed list of initial and future fleet. Quotes on all major equipment.
Plant capital cost estimate basis: • Estimate prepared by experience, factor, unit cost, shelter volume, etc. • Estimate form—generally none. • Site visit by Owner's estimator. • Civil/structural—unit volumes from sketches. • Architectural—historic factors. • Mechanical—percent of machinery. • Piping/HVAC—percent of machinery. • Electrical—historic; e.g., unit cost/hp or per kW, or percentage of total cost. • Labor rates—historic factors. • Productivity—not evaluated. • Material volumes—historic figures with a percentage factor for installation. • Material/equipment pricing—either from manuals or historical data bank. • IT and control systems—percent of budget. • Assume prices FOB site, with tax.	**Plant capital cost** estimate basis: • Estimate by probable unit cost or spread (including installation). Key elements are all bottom up, task based; but many items still top down. • Estimate form—desirable. • Site visit by project estimator—one letter quote for each major equipment. • Civil/structural—take-offs from sketches for major items. • Architectural—partial take-offs. • Mechanical—preliminary take-offs. • Piping/HVAC—take-offs on major items. Factors on rest. • Electrical—area rates for most items; historical factors elsewhere. • Labor rates—known, generic/union construction rates from region. • Labor productivity—assumed. • Construction equipment usage— percent of labor. • Material volumes/amounts—take-offs on major items only. • Bulk commodities from initial GA drawings and sketches. • Material/equipment prices—from EPCM; a few single vendor quotes. • Major long-lead-time items known. • Spares list and initial fills factored. • IT and control systems—percent of budget. • All prices FOB site, include tax.	**Plant capital cost** estimate basis: • Estimate is now mostly all bottom up from sound engineering criteria, equipment list, and take-offs. • Estimate form—generally required. • Site visit by estimator and major equipment vendors—two quotes. • Civil/structural—partial take-offs and quotes, benchmarked to real data. • Architectural—mostly take-offs. • Mechanical—mostly full take-offs. • Piping/HVAC—partial take-offs and hours, benchmarked to real data. • Electrical—present rates. Take-offs and hours for electrical runs. • Labor rates—actual construction rates and burdens for the area. • Labor productivity—evaluated. • Construction equipment usage—percentage of labor/crew. • Material volumes—partial take-offs; rest by weight factor or experience. • Bulk commodities quantified from plot plans, GAs, and design drawings. • Material/equipment pricing—detail list using quotes from multiple vendors. • All long-lead-time items identified. • Spares list and initial fills estimated. • IT and control systems—calculated. • All prices FOB site, include tax.

Figure 4.2 Comparison of project stage work efforts (continued)

SCOPING EVALUATION	PREFEASIBILITY STUDY	FEASIBILITY STUDY
Infrastructure: Facilities costed from historic factors.	**Infrastructure**: Facilities costed from quotes or estimating manuals.	**Infrastructure**: Costed from formal written quotes on all major facilities.
Contractors' OH&P: Included in unit cost or set as percent of total cost.	**Contractors' OH&P**: Percent of direct cost with factors for indirects. • Subcontract pricing—historical.	**Contractors' OH&P**: Written quotes from qualified bidders. • Subcontracts—written quotes.
EPCM: Percent of total budget.	**EPCM**: Percent of direct cost.	**EPCM**: Estimate from EPCM firm.
Prefeasibility and feasibility study: Cost as a percentage of total cost.	**Feasibility study**: Cost by firm quote.	
Owner cost/pre-production expense: • Generic top-down estimate or a percentage of total budget. • Corporate overheads excluded.	**Owner cost**/pre-production expense: • Mostly factoring of similar projects; bottom-up estimates for key items. • Sustaining capital shown. • Corporate overheads excluded.	**Owner cost**/pre-production expense: • Zero-based budget calculation. • Sustaining capital shown. • Corporate overheads excluded.
Escalation: Generally excluded.	**Escalation**: Use Owner's allocation.	**Escalation**: Define by area and risk.
Working capital: • Factored from Owner methodology. **Exclusions** • Not applicable.	**Working capital**: • Estimate prepared by experience factor from similar project. **Exclusions**: • Broadly stated.	**Working capital**: • Detailed estimate calculated from zero-based budget by Owner. **Exclusions**: • List of items agreed to by Owner. • Needs to be kept to a minimum.
Contingency: • Generic estimate. Either set as a single percentage, or heavy factors utilized to cover the unknowns. • +40% overall; range +30% to +70%. • Equipment +15% to +25%. • Structures +20% to +50%.	**Contingency**: • Broad evaluation. Overall project contingency still heavy. Many items still estimates. • +22% overall; range +15% to +30%. • Quoted equipment +10% to +15%. • Structures +15% to +20%.	**Contingency**: • Detail engineering and zero-based budgeting should reduce overall contingency to <18%. • +15% overall; range +12% to +18%. • Typically no single area >20%. • Equipment with firm quotes +5%.
Accuracy at 90% confidence level: • Expect ±40%; range +100% to −50% for overall project. • Accuracy judgmentally assessed.	**Accuracy** at 90% confidence level: • Expect ±25%; range +40% to −30% for overall project. Not "bankable" quality. • +30% to −20% for individual facility. • ±30% for site closure plan. • Accuracy statistically evaluated.	**Accuracy** at 90% confidence level: • Must be ±15% or better. (Historical range of +25 to −20% is not definitive.) • +20% to −10% for individual facility. • ±20% for site closure plan. • Monte Carlo simulation to yield <10% probability of overrun, i.e., 90% confidence project cost will be less than or equal to budget.

Figure 4.2 Comparison of project stage work efforts (continued)

SCOPING EVALUATION	PREFEASIBILITY STUDY	FEASIBILITY STUDY
OPERATING COST ESTIMATE		
Operating quantities: • General, i.e., generic historic. • Workforce estimate.	**Operating quantities:** • Estimates (some factoring). Mix of bottom-up and top-down procedures. • Personnel list—by department.	**Operating quantities:** • Detailed; mostly bottom up. Minimal factoring (chiefly for maintenance). • Personnel list—by job function.
Mine operating costs: • "Top down," order of magnitude.	**Mine operating costs:** • Preliminary estimate from manuals and benchmark of other properties.	**Mine operating costs:** • Zero-based budget. Vendor input. Breakdown by mining function and year.
Plant operating costs: • Collection and usage of any existing cost data. Order-of-magnitude estimates from manuals and data banks used where no data exists.	**Plant operating costs:** • Labor, supplies, fuel, power, reagents, and liner costs, etc., chosen by benchmark and factoring of similar projects, or from estimating manuals.	**Plant operating costs:** • Detailed from zero based budget. • Formal quotes for major consumables. • Labor costs (rates and burdens) calculated from review of similar actual local area data and contracts.
• Published regional power costs. • Benchmark overall unit costs utilized. • Operating costs shown yearly in total dollars and per unit produced.	• Maintenance costs are estimated as a percentage. • Power costs from local utility contacts. • Costs broken out by work area. • Operating costs shown yearly in dollars, per ton, and per unit produced.	• Letter quote from vendors for most supplies. Minimal historical data. • Written proposal from local utility. • Costs by area and element. • Operating costs shown yearly in dollars, per ton, and per unit produced.
Freight, smelting, and refining costs: • Generic from manuals. **Administrative costs:** • Generic. **Insurance:** • Order-of-magnitude factor.	**Freight, smelting, and refining costs:** • Estimate from preliminary quotes. **Administrative costs:** • Owner estimates (some factoring). **Insurance:** • Historical factor.	**Freight, smelting, and refining costs:** • Detailed from submitted bids. **Administrative costs:** • Detailed, using sizable Owner input. **Insurance:** • Budget quotes from Owner.
Accuracy: • Likely to be no better than ±33%.	**Accuracy:** • Typically +25% to –10%.	**Accuracy:** • Typically +15% to –5%.

Figure 4.2 Comparison of project stage work efforts (continued)

SCOPING EVALUATION	PREFEASIBILITY STUDY	FEASIBILITY STUDY
MARKET PLAN AND MARKET POTENTIAL		
• Marketability of product discussed. • Very high level analysis; no depth.	• Market size and demand estimated.	• Marketing plan derived by third party. • Identification of potential buyers.
Financial analysis: • To demonstrate broad viability. • PEA. • Royalties and taxes—ignored. • Mineral prices from trade journals. • Simple spot cash flow analysis. • Preliminary pre-tax IRR. • Preliminary $NPV1_0$. • No sensitivity analyses.	**Financial analysis:** • Inputs OK for reliable results. • Assess key economic parameters. • Royalties and taxes—estimated. • Mineral prices set. • Cash flow analysis by Owner. • Unit cash cost and full cost derived. • List of all assumptions made. • Pre-tax IRR. • Likely tax issues noted. • Post-tax IRR optional. • NPV at Owner discount rate. • One or two sensitivity analyses.	**Financial analysis:** • The final economic decision tool. • Full assessment of all parameters. • Royalties and taxes—Owner derived. • Mineral prices set. • Sophisticated cash flow analysis. • Unit cash cost and full cost derived. • Comprehensive list of all assumptions. • Pre-tax IRR. • Full investigation of tax regime. • Post-tax IRR, ROI, and payback. • NPV at Owner discount rate. • Multiple sensitivity analyses.
BUSINESS CONSIDERATIONS		
	• Highlight any competitive advantage.	• Highlight any competitive advantage.
• State the potential value to Owner.	• List rewards project brings Owner.	• List rewards project brings Owner.
	• Reconciliation of the project's returns with corporate objectives.	• Reconciliation of the project's returns with corporate objectives.

Figure 4.2 Comparison of project stage work efforts (continued)

SCOPING EVALUATION	PREFEASIBILITY STUDY	FEASIBILITY STUDY
FINANCING		
• Usually not addressed.	• Preliminary discussions to establish availability of funding/insurance.	• Scoping study reviewing availability of financing and insurance monies.
	• Foreign exchange strategy—discuss.	• Foreign exchange strategy—defined.
	• Preliminary currency risk assessed.	• Currency exposure to nearest $5 million determined.
		• Owner's treasury department determines financing and interest payments during construction.

Notes: 2-D = two dimensional; 3-D = three dimensional; ARD = acid rock drainage; ASTM = American Society for Testing and Materials; CADD = computer-aided design and drafting; CAPEX = capital expenditure; COG = cutoff grade; CPM = critical path methodology; EA = environmental assessment; EIS = environmental impact statement; EPCM = engineering, procurement, and construction; ESIA = environmental and social impact assessment; ETAP = electrical transient analyzer program; FOB = free on board; GA = general arrangement drawing; GMP = guaranteed maximum price; HAZMAT = hazardous materials; HAZOP = hazard and operability; hp = horsepower; HVAC = heating, ventilation, and air conditioning; IRR = internal rate of return; IT = informational technology; JORC = Joint Ore Reserves Committee; m = meter; MTO = material takeoff; NI = National Instrument; NPV = net present value; OH&P = overhead and profit; P&ID = piping and instrumentation diagram; PEA = Preliminary Economic Assessment; PEP = Project Execution Plan; PFD = process flow diagram; QA/QC = quality assurance and quality control; RC = reverse circulation; RFQ = request for quotation; ROI = return on investment;
SEC = Securities and Exchange Commission; SG = specific gravity; T&M = time and materials; TSF = tailings storage facility; WAN = wide area network; and WBS = work breakdown structure.
v
The tabulation lists typical minimum ingredients for the four report phases of a major greenfield project. Not all elements are needed on small projects. This is a guide; not a list of mandatory items for every project.

Figure 4.2 Comparison of project stage work efforts (continued)

SECTION 5

Management Review, Goverance, Stage Gate

OBJECTIVE

Once the Scoping Evaluation, and its associated components, is completed, the next step in the stage gate process is a full review with senior management. This review should be comprehensive regarding the results of the Opportunity Stage Scoping Evaluation and the next steps required to move forward to the viability stage.

Each section of these guidelines is designed to direct the reader to the corresponding chapter in *Project Management for Mining* (*PMfM*), which must be thoroughly understood for successful implementation. *PMfM* must be used alongside these guidelines.

REFERENCES

Item	*PMfM*	Page
Scoping Evaluation	Chapter 5	49
Management Review	Chapter 6	61
Governance	Chapter 4	33
Ethics	Chapter 3	23

CG Figure	Figure Caption	*PMfM* Reference
Figure 5.1	Governance program	Figure 4.4, p.42
Figure 5.2	Management review project approval process	Flowchart 6.1, p.66
Figure 5.3	Authorization for expenditure	Flowchart 16.1, p.270
Figure 5.4	Example Authorization for Expenditure form	New Figure

Checklist Title	*PMfM* Reference
Opportunity and viability	New checklist
Stage gate final AFE before execution	New checklist

SCOPE

The PM needs to have clear documentation of the opportunity in order to present it to senior management. The approach is similar at each stage of project development and must be carried out according to company policy and project management best practices.

Management Review

The primary objective is to formally apprise management of the project potential.

Scoping Evaluation

A fully documented Scoping Evaluation and economic analysis is completed (*PMfM* Flowchart 5.1, p.60).

Operating Team Review

The results of the Scoping Evaluation should be reviewed with a team of operations management.

Management Review Submission

Preparation for the formal review includes the list of materials shown in Figure 5.2. A review of *PMfM* Chapter 5 is also recommended.

Management Review and Decision Process

The management review process should involve a discussion about next steps. If the Scoping Evaluation was completed in-house, then the discussion should include having a third-party project delivery expert review the entire project and conduct a gap analysis before moving to the Viability Stage.

Opportunity to Viability Stage Steps

Once the review is complete, the next steps are outlined in Figure 5.2.

Attributes of Good Governance

Governance is not a project management function or activity, nor is it an additional layer of decision-making. Good governance should consider all of the following to ensure alignment and objective evaluation and development of projects:

- An explicit organizational structure that enhances accountability.
- The governance structure encourages adherence to designated authority.
- The structure sustains fulfillment of corporate commitments.
- The structure enables checkpoints and fosters communication lines.
- Processes are not mere formalities to mark compliance.
- An objective monitoring process that can impartially assess investment readiness, facts, and data driven decision-making.

- Project performance is transparent.
- A series of go/no-go executive decisions over the course of the project life cycle. The project is not allowed to move around the gate (approval point) processes.
- Third-party independent reviews enhance objectivity and eliminate optimism bias.
- An information-sharing process that streamlines and empowers decision-making. Deliverables are completed collaboratively.
- The risk of project surprises is reduced, and quality is enhanced.
- Bureaucracy is minimized by developing templates of required reports and ensuring required content is identified.
- A project team that shares a holistic view of project success. The governance oversight entity and the project team are cohesively aligned.
- There is cross-functional alignment of key business and technical requisites.

Stage Gate

The review process that takes place at this juncture (commonly known as a *stage gate*) is the first of a series of approval steps that the project will have to pass through if it is to become a viable project. At each gate, the continuation of the project is decided by a senior company authority. The decision to go forward, or not, is based on the information available at the time, including the business case, risk analysis, and availability of the requisite resources (e.g., money and people with the correct competencies). This initial management review is a part of project governance, as described in *PMfM* Chapter 4.

Figure 5.1 portrays how the governance function and its go/no-go decision gates relate to the project management function.

Management Review Components

At each stage gate from Scoping through FS, there are critical elements of the review process. Components include:

- The study (Scoping Evaluation, PFS, FS)
- Total Project Cost

Total Project Cost (*Companion Guidebook (CG)* Section 9) provides all funding required from the end of FS through the execution stage (engineering, procurement, construction, commissioning, start-up, and ramp-up to commercial production). Commercial production is defined as operating on a cost-neutral basis for one calendar month. The Total Project Cost must be included in the economic analysis.

It is a fact that over 85% of major and mega projects go over budget during execution. For executives concerned about budget overruns, it is recommended that their corporate financial department establish, early in the execution stage, a separate Management Reserve (i.e., completely unavailable to the PM) of up to 35% more than the calculated Total Project Cost. Management Reserve is defined in *CG* Section 9.

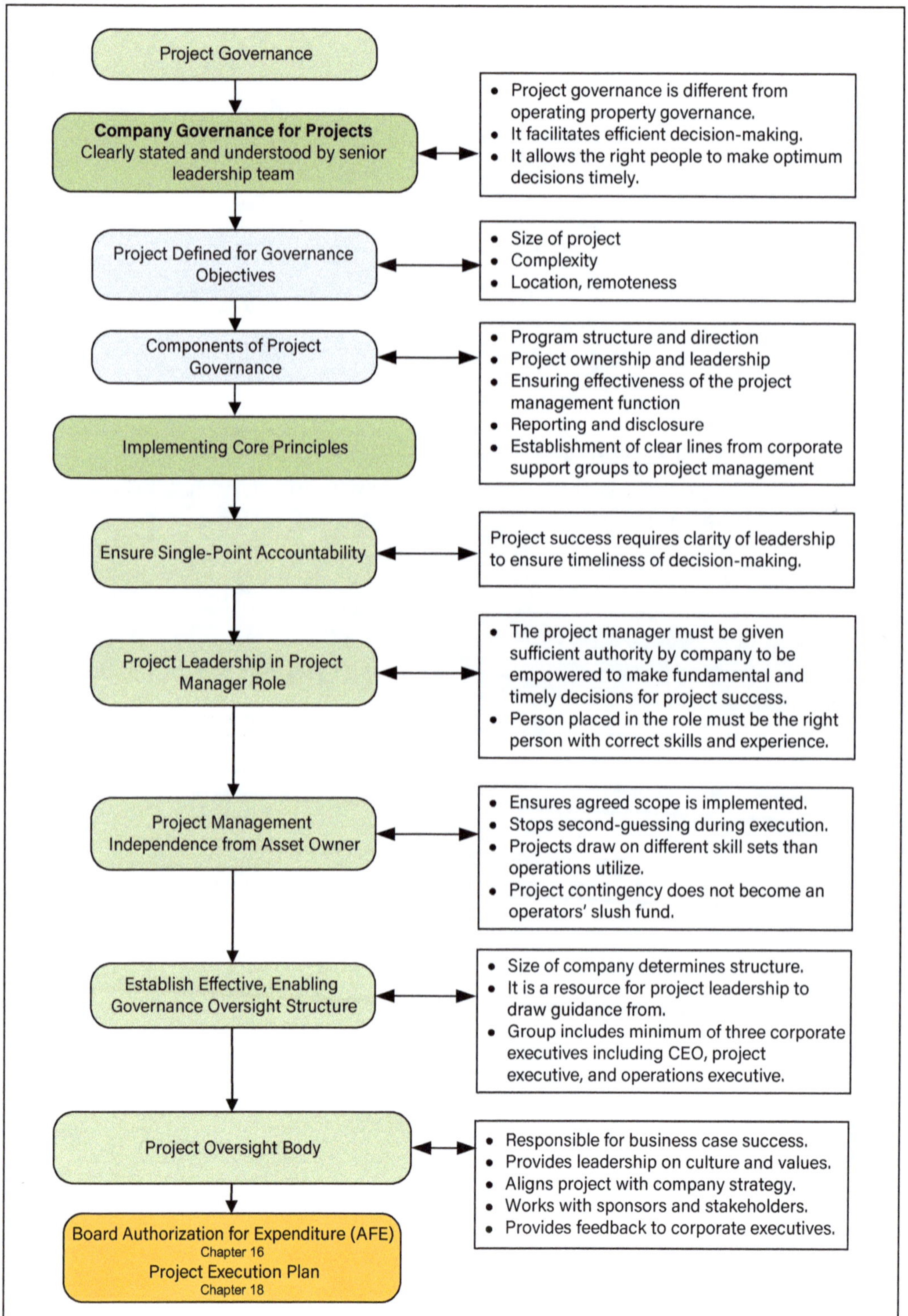

Figure 5.1: Governance program

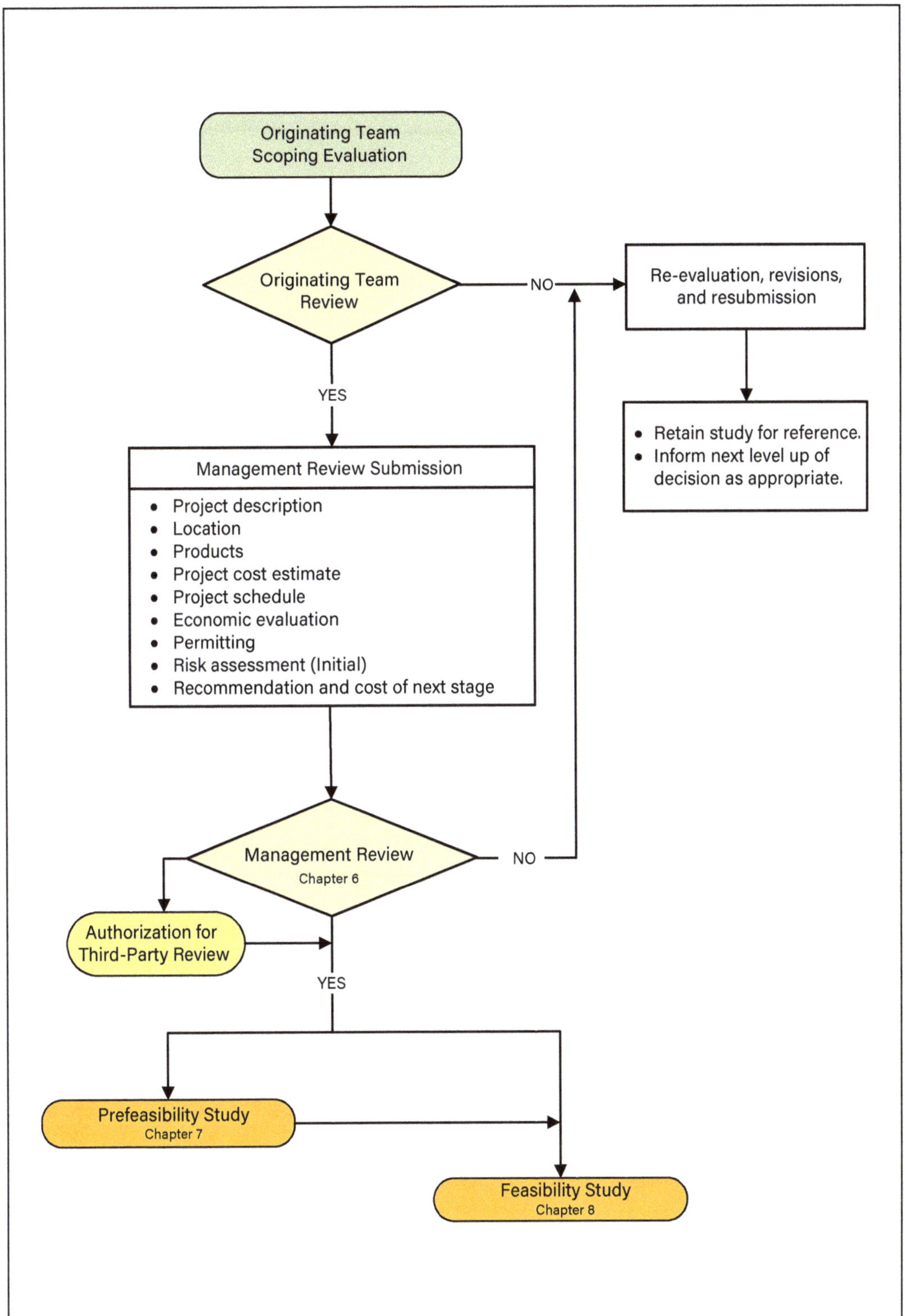

Figure 5.2: Management review project approval process

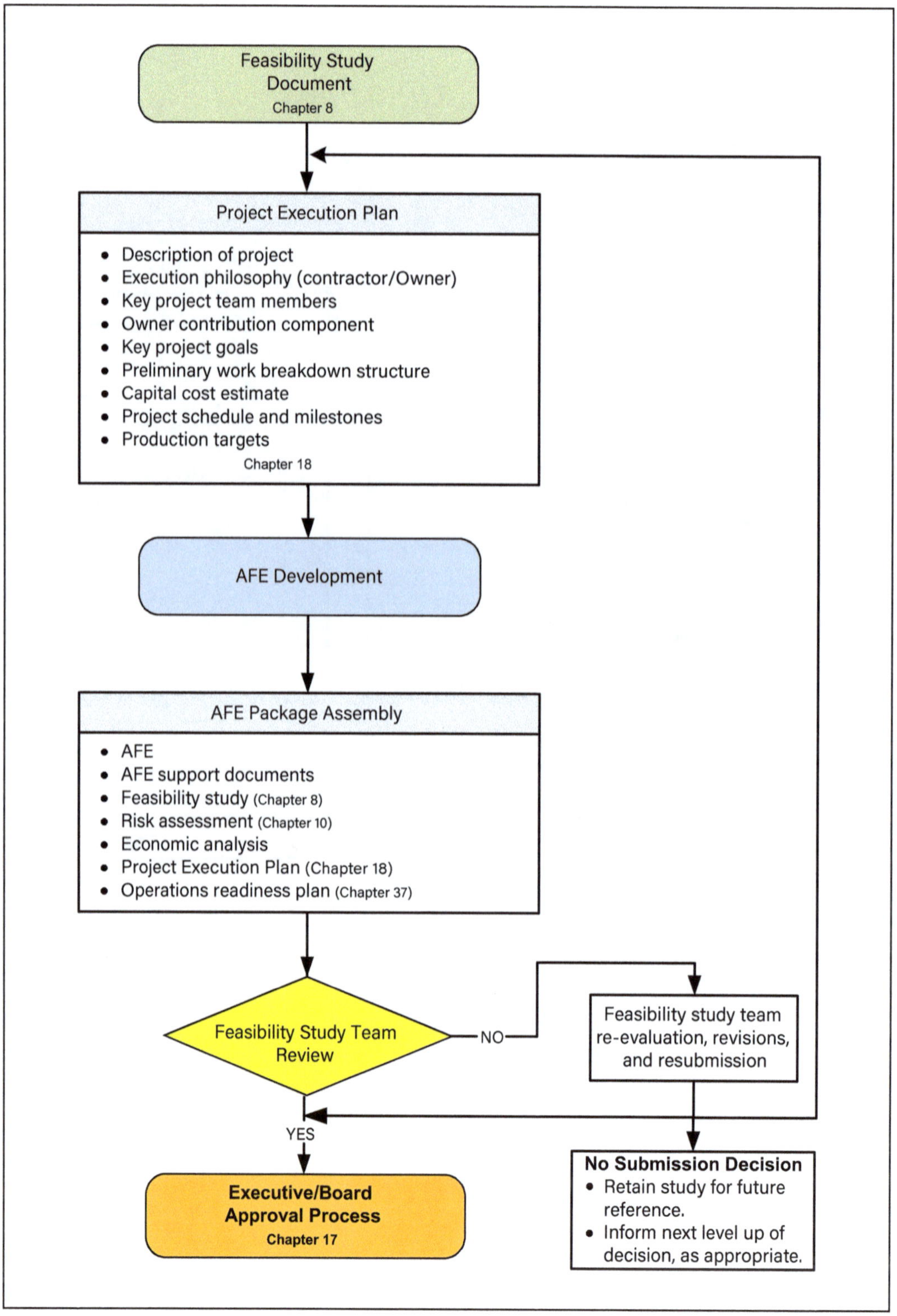

Figure 5.3: Authorization for expenditure

Company Name
Authorization for Expenditure (AFE)

Project Title ____________ Group/Division **name**
Project Manager ____________ initials
Area Manager ____________ initials

Project Timing and Identification	*Authorization Requested*	*Allocation Timing*
Start Date ______	Capital ______	**Year** **Amount**
Completion Date ______	Lease ______	xx year(s)
AFE Type ______	Expense ______	Beyond
AFE Number ______		Total $ - USD
Contractor Ref ______		
Prepared By ______		

Project Description: (Provide a summary description here. NPV analysis, quotes and other details should be attached in addition sheets)

Additional details and supporting documention attached:

Financial Results	***Key Project Aspects*** (All PA & PM except replacement in kind) *List major advantages, challenges and risks associated with the project, incl. magnitude, timing and duration. Discuss sustainable competitive advantage.*
IRR ______	
ROR ______	**RISKS** -
Repatriated IRR ______ (if applicable)	**OPPORTUNITIES** -
NPV ______ Payback (years) ______	**JUSTIFICATION** -

Review Signatures - *Confirm accuracy, relevance, completeness of data, analysis assumptions related to your functional area or management contribution to this project. Comments to assist the Approver(s) should be attached to this document.*	***Approval Signatures***
Name **Function Date**	Signature Date
	VP Technical Services Name and Title
	Signature Date
	CEO Name and Title
	Signature Date
	Name and Title

Figure 5.4: Example Authorization for Expenditure form

Checklists: The stage gate requirements including AFE documents are different between Opportunity, Viability, and Execution stages.

OPPORTUNITY & VIABILITY

No.	Item	Status	Date
1	Scoping evaluation analysis—complete		
2	Owner's organization review level—determined		
3	Owner's management review submission—prepared		
	A. Concise project description—completed		
	B. Project scope of work, going forward—outlined		
	C. Development schedule—estimated		
	D. Project capital cost—estimated		
	E. Project economics (pretax IRR and NPV)—derived		
	F. Future operations production and unit costs—estimated		
	G. Key assumptions—documented		
	H. Significant risks—identified		
	I. Scoping study conclusions—summarized		
	J. Reconciliation of project benefits and deliverables with owner's corporate objectives—produced		
4	Recommendation for prefeasibility or feasibility study—determined		
5	Recommendation for project management of next stage—resolved		
6	AFE for scoping, prefeasibility or feasibility study—prepared		
	A. AFE guidelines review—undertaken		
	B. AFE form—prepared, signed, approved		

STAGEGATE FINAL AFE BEFORE EXECUTION

No.	Item	Status	Date
1	AFE execution stage package—prepared		
	A. Feasibility study—complete		
	B. Preliminary project execution plan—complete		
	C. Operations readiness plan synopsis—complete		
2	Owner's organization review level—determined		
3	Key project stakeholders—identified		
4	AFE package and supporting documents—submitted		
	A. Project memorandum—prepared		
	B. Executive summary—included		
	C. Capital budget expenditure summary—included		
	D. Project execution schedule—included		
	E. Future operations production schedule—included		
	F. Economic analysis—produced		
	G. Project assumptions, risks, and value drivers—documented		
	H. Key project benefits, deliverables, and goals—highlighted		
5	Originating team approval—submitted		
6	Owner's organization—initial senior management level review approval		
7	Owner's president and CEO—executive approval		
8	Owner's Board of Directors—final approval		
9	Original AFE and approvals—filed with controller and project team		

SECTION 6

Opportunity Stage Advanced Exploration: Idea To First Study

OBJECTIVE

The Opportunity Stage identifies the steps required to evaluate a potentially viable idea. This section of the *Companion Guidebook (CG)* relates to moving an advanced exploration target, or other valid idea, through a successful first Scoping Evaluation.

Each section of these guidelines is designed to direct the reader to the corresponding chapter in *Project Management for Mining (PMfM)*, which must be thoroughly understood for successful implementation. *PMfM* must be used alongside these guidelines.

REFERENCES

Item	*PMfM*	Page
Scoping Evaluation	Chapter 5	49
Project Stage Work Efforts	Chapter 15	243

CG Figure	Figure Caption	*PMfM* Reference
Figure 6.1	Scoping evaluation	Flowchart 5.1, p.60

Checklist Title	*PMfM* Reference
Scoping evaluation	Checklist 5.1, p.58 (modified)

SCOPE

Scoping Evaluation and Idea Development

The primary goal of the Scoping Evaluation is to ensure that the assessment of an idea or advanced exploration target includes all the major facets required to develop into a potential project and reaches a valid economic conclusion.

- *PMfM* Chapter 5 identifies the steps that follow the identification of a potentially viable idea and then the development of that idea into an opportunity capable of being presented to senior management within a Scoping Evaluation document. This is the document that allows for either rejection or further pursuit (*CG* Section 6). The Scoping Evaluation may also be known as a scoping study or a conceptual study.
- In some public disclosure formats, a PEA can be promoted to serve as the first evaluation. However, the NI 43-101[1] PEA does not meet the best -practices standards as defined in *PMfM*.

Advanced Exploration Mineral Resource Evaluation Challenges

It must be understood that there is not an intermediate step to a valid Scoping Evaluation. The first characterization of whether the mineral resource can be economic is a reliable Scoping Evaluation.

The challenges a team advancing a promising exploration target faces are complex, requiring exploration teams to work on multiple fronts. Advanced exploration moves into the gray area of what is a project, but the process is itself a project; there is a scope, data, processes, and expected outcomes.

As the focus shifts for a promising advanced target, a question needs to be answered: What does it take to make this mineral resource economic?

- What is the potential size of the resource based on a reasonable level of classification?
- What is the potential life of mine, and will this cover the investment and provide profitable years?
- What is the potential upside of the deposit or district?

The mineral exploration objective of identifying sufficient mineral resource to support an economically viable mine must always have the "end goal" in mind. The first "end goal" is to deliver a Scoping Evaluation with sufficient reliability for a decision to be made either to pursue or reject.

Scoping Evaluation Activity

It is necessary to put the completed evaluation into a format that is consistent with the project delivery process and can serve as a foundation for the next steps in the development process if the project is approved to move forward to the Viability Stage. The activity must therefore follow the guidelines below (Figure 6.1).

Select Team

Recognizing that people without project experience generally lead this effort, the team leader for this study should be someone with sufficient authority to work with all the in-house organizations and who is able to work to deadlines.

Data Collection and Analysis

- Free exchange of all related data within the mining organization is necessary to ensure all the key points are analyzed during the evaluation. Today, data is assembled in complex models that require understanding by experts in their fields.
- It is critical to collect key information during exploration activity to reduce the costs and time required to advance the project.

Concept Development

- Ensure the project scope for all areas is clearly defined.
- The conceptual developments must be done by an engineering group, internal or external, to assist with the very preliminary plans that must be put together.

Permitting, Environmental, and Social

Understanding the relevant requirements will assist in the project timeline and overall project cost. Early discussions in these areas will help minimize confusion and delay later.

Risk Evaluation

The initial risk evaluation should be undertaken during the Scoping Evaluation.

Capital and Operating Cost Estimate

The accuracy of the Scoping Evaluation is generally in the ±40% range. That accuracy can be greatly affected by the amount of preliminary design work completed, allowing for a better-defined scope.

Level of Effort

The work effort required to be completed for a Scoping Evaluation is defined in *PMfM* Figure 15.2 and shown in *CG* Section 4.

Economic Evaluation

The economic evaluation should be done by, or with, someone from the corporate financial group who understands the key drivers for the size and impact of the opportunity.

Scoping Evaluation and Recommendation

- Besides completing the Scoping Evaluation, a set of recommendations should be presented to senior management related to next steps for the opportunity, along with a budget and schedule (*CG* Section 5).
- The key purpose is to allow a management decision to reject or pursue.
- The decision to pursue advances the project to PFS or FS, with a gap analysis conducted to confirm the scope of work going forward.
- Exploration may need to continue, with certain focus areas defined, in order to complete a follow-up Scoping Evaluation or decide to discontinue exploration activity.
- The decision to reject should be clear once the economic evaluation is complete.

Project Manager must make certain all aspects of Opportunity are addressed.

- Where no data exists, judgment calls and best estimates are made.
- A preliminary economic evaluation must be generated.
- To manage the uncertainties in early-stage opportunities, the caliber and experience of team personnel must be superior.

Ethics and corporate policies must be considered by the Advocate and PM (*PMfM* Chapters 3 and 4). The PM is responsible for ensuring there is no violation of company principles or policies.

Areas to Be Considered for Scoping Level Evaluation

The social, community, and economic impacts of the project need to be considered early. The exploration and project teams need to consider the following topics:

Societal Considerations

- Developing a social, environmental, and community development strategy (*CG* Section 10) up front is critical in today's environment. Continue to build on exploration group relationships.
- Sustainability and the social acceptance to operate are critical; do not treat this as a government "must do." It is the mining company that must take the lead.
- Exploration is often the first team on the ground interacting with local communities and government. Building relationships inline with company principles is key. Building honest relationships with appropriate expectations will improve project timing and success. "Trust gained translates to speed gaining agreements" in the case of community relationships.
- Obtain the relevant permits, including all social acceptance processes. This is the beginning of a potential multi-decade community relationship. The goal is a continuing license to proceed:
  - Find out whether you must negotiate each time you need a community decision.
  - Begin the relationship with the community immediately. The sooner, the better. Do not wait until you have all the information before starting work with them.
  - Think outside your business. Start by supporting their existing activities.
  - Source and hire locally. Set targets and plans to achieve those targets, including some basic training. This is low cost, and it will add great value. Find a role for everyone in the community, if possible.
  - Find local entrepreneurs and work with them to develop allies. See your project and future mine as a potential market for them.
  - Be aware of, and honest about, how this potential project may impact local lives.
- Determine the environmental considerations in the country of exploration. It is critical to understand this at the outset.
- Baseline data is key and takes many seasons. The more data you have, the easier it is to make the case for environmental offsets in the management plan.
- Involve a seasoned social specialist. This is also the start of capturing social data and informing local communities about the company's activities.

Property

- Exploration will need land access to get drills to the mineral targets. If an area starts to appear strongly positive for a commercial deposit in the scoping stage, it may make sense to purchase it outright rather than waiting for the Definitive feasibility. The more negotiations you have, the longer the ensuing project schedule.
  - The project will likely find the lowest land prices during the exploration stage.
  - A possible rule of thumb is to pay up to 1% of the potential NPV of the operating mine at this point.
- Identify the position, ownership, and control of land and other business issues.
- Are properties and land owned by a community? The project needs to understand community governance (i.e., the decision-making process).
- On-site exploration staff can help a project by determining whether utilities (power, gas, water) cross multiple regions, communities, government jurisdictions, or private land.
- Would there be a benefit to upgrading the current exploration road access early to a more permanent road for developing and operating a mine? Or does road building lead to population influx and negative impacts such as deforestation and non-company habitat impact?

Water

During initial project review or during exploration is a good time to determine how to secure control of water for any future project. Understanding the political issues related to water supply and quality will assist with the project scope and opportunities for community involvement and projects.

Required Advanced Exploration Level of Detail

Advancing a mineral resource to a valid scoping level requires the exploration team to focus on areas beyond drilling, often requiring support outside of the Exploration Group.

- The mineral resource development must advance to the appropriate level of reliability.
- Determining the quantity of mineral resource at the observed grade is required to support a viable mine plan.
- Answer the question: Is diamond drilling (DD) or reverse-circulation drilling (RC) more appropriate for this deposit? Is it all DD, or can some of the deposit be drilled with RC?
  - The more complex the deposit, the more critical it is that DD provides >70% of the data.

- Overuse of RC can be problematic for acceptable mineral resource determination (type of deposit, mineralization, and other controls will dictate what information is required).
- Preservation of drill samples in the original configuration (correct orientation, lack of core breakage from rough handling, etc.) is vital in the on-site core management process for future data reconciliation and auditing.
  - Digital scans as DD emerges, and other digital analyzers to better define minerals, can aid in resource confidence.
  - Rock Quality Data (RQD) and Rock Mass Rating (RMR) must be reviewed very close to the drill hole so transport does not degrade DD samples (digital imaging and/or downhole televiewer augmentation will all aid in data credibility).
- Determine early the kind of data needed; then collect as much as possible while drilling is ongoing to reduce later redrilling time and cost.

- What, if any, test work can be conducted? How much DD or RC sample is required?
  - Define what minimum percentage of evidence from DD and RC samples is required. Can this be reduced with quality scanning or imaging?
- Define the data models required and advance the modeling using preliminary lab results or field logging and estimation.
- Establish the Quality Assurance / Quality Control (QA/QC) protocols to follow.
  - Do not overlook multi-element or other assay suites that may be useful, including deleterious elements in processing, economic mineral credits, vectors (geochemical signatures), mineral domains, and environmental concerns.
  - Exploration teams often require fast turnaround on samples and may prioritize quick results over chain-of-custody and QA/QC procedures. Ensure that a comprehensive Data Quality Management system is in place and supports both rapid turnaround and best practices around data capture and management.
- Consider the company's preferred approach to securing the mining rights, business plan, and project controls.
- Understand commercial issues such as the terms for processed minerals and in-country beneficiation rules.

Models to be Supported by Exploration Results for Scoping Evaluation

No.	Model	Description
1	Geological	Fundamental model supporting other models.
2	Resource	Provides an appropriate estimation of grade and distribution based on study stage (including any deleterious elements).
3	Geometallurgical (Metallurgy)	Interpretations for ore zone development, scoping level process concepts, and initial estimates on product quality for marketing purposes supported by geochemistry and mineralogy.
4	Hydrogeological	May fit under geotechnical scope for mine water issues; or may be across the project for complex water supply, discharge, and water balance management.
5	Geotechnical	Initial geotechnical findings for scoping level mine concepts, future mine planning, as well as fragmentation and comminution considerations.
6	Mine Planning	Initial mine planning model supported by the various models listed as required by the complexity and level of study.

Additional Resources Required to Support Advanced Exploration Team

If a first pass project scope is to emerge from advanced exploration, the exploration group will need help from project group resources (e.g., mining, processing, legal, environmental, and marketing teams or personnel) that can help define what the mineral resource will be expected to support:

- Is the project likely to be an underground or open pit mine?
- What are the process options for recovering the economic mineral(s)?
- What are the constraints in developing this mineral resource: legal, environmental, sustainability, country factors, etc.?
- What market factors should the exploration group keep in mind, i.e., metal or mineral price, demand for final product, the quality required for sale, etc.?
- Infrastructure requirements: supply lines, labor availability, utilities, etc.
- Social investigation and how best to involve local communities to gain acceptance of mine development in their area.

The project group resource should be chosen to provide judgment calls and best estimates where no data exists. Therefore, the caliber and experience of project support must be significant (generally >10 years).

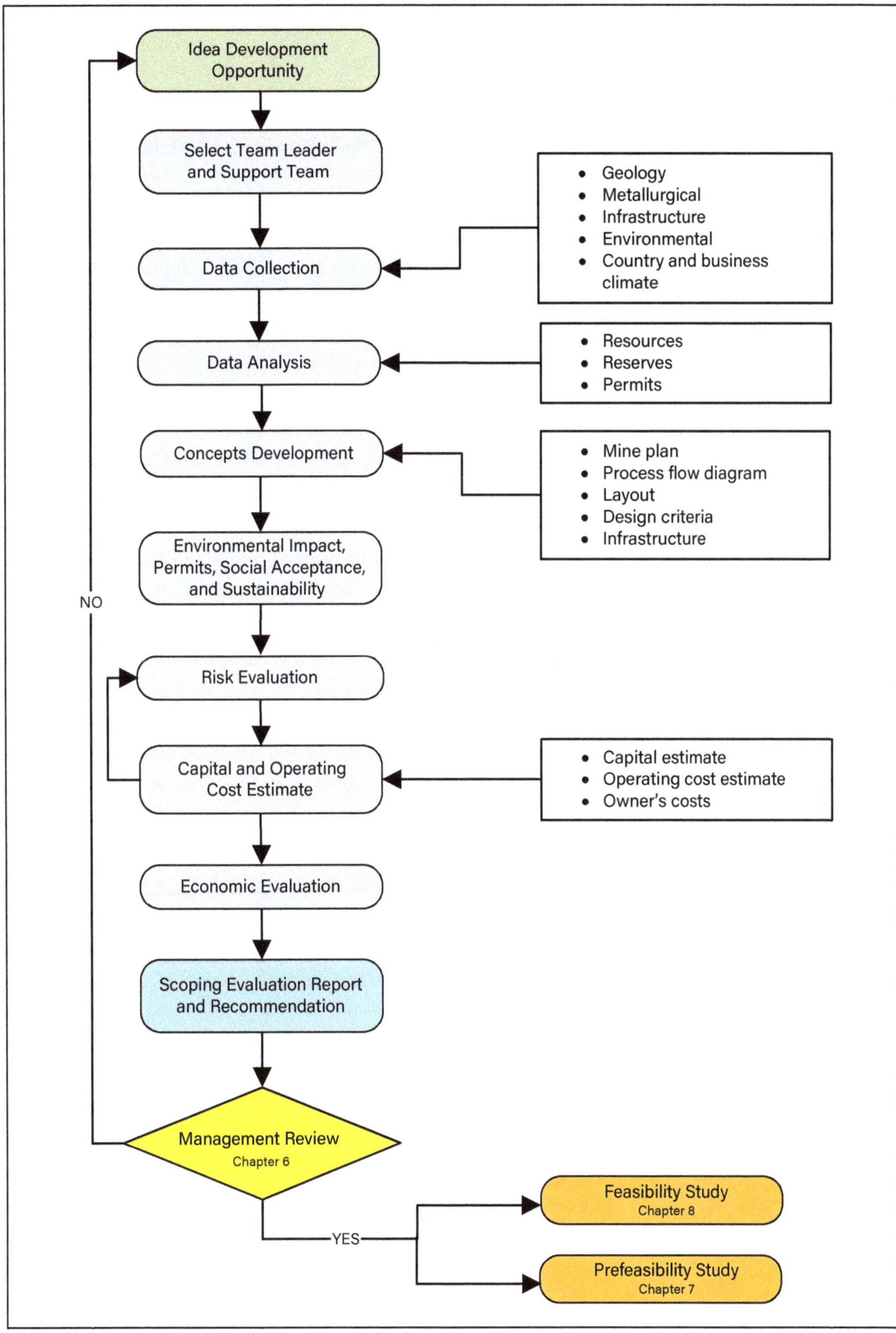

Figure 6.1: Scoping evaluation

SCOPING EVALUATION

No.	Item	Status	Date
1	Target identified		
2	Advanced exploration—idea formally characterized as an opportunity		
3	Management consensus to pursue opportunity		
4	Agreed-upon project scope with management		
5	Define work for scoping evaluation, i.e., set the deliverables (Reference the first column in *PMfM* Figure 15.2)		
6	Identify critical elements of work that must be completed		
7	End result deliverables defined up front		
8	Name project leader		
9	Identify project team—in house and / or outsource consultants		
10	Estimate cost of evaluation		
11	Determine property ownership status		
12	Select engineering support contractor—if applicable		
13	Initiate scoping evaluation study		
14	Site visits by:		
	A. Project leader		
	B. Company business opportunity development person		
	C. Technical support specialists—e.g., Mining / metallurgical		
15	Exploration geology program:		
	A. Geologist—site visit		
	B. Geologic assessment and sampling—undertaken		
	C. Verification of existing drill hole data		
	D. Drilling—optional—as appropriate		
	E. Geological resource—estimated		
16	Mining program:		
	A. Preliminary reserve (tonnes and grade)—estimated		
	B. Mining method—assigned		
	C. Production schedule for mine life—estimated		
17	Models to be developed:		
	A. Geologic modeling		
	B. Geometallurgy interpretations for ore zone development, and scoping level process concepts		
	C. Initial geotechnical findings for scoping level mine concepts and future mine planning		
	D. Hydrogeological model		
	E. Metallurgical lab tests on collected samples—optional		

SCOPING EVALUATION (continued)

No.	Item	Status	Date
18	Engineering design:		
	A. Conceptual engineering—undertaken		
	B. Flow sheet—derived		
	C. Design concept—outline		
	D. Site location map		
	E. Major equipment list		
	F. General process description		
19	Infrastructure / support facilities—described		
20	Hydrology—discussion		
21	Power—overview		
22	Logistics—review		
23	Supply base availability—review		
24	Work force availability—review		
25	Environmental studies:		
	A. Environmental specialist—site visit		
	B. Overview of major environmental considerations and risks		
26	Permitting needs and permitting constraints—identified		
27	Project master schedule (showing construction start and end)		
28	Gantt chart of major work element durations		
29	Monthly progress reports—published		
30	Capital cost estimate (accuracy of –50 to + 100%)		
31	Operating cost estimate—overall, by unit cost		
32	Contingency calculation—appropriateness confirmed		

ENDNOTE

[1] National Instrument 43-101 and Canadian Institute of Mining standards are written for public disclosure on the Canadian stock market. JORC, Sk13000, and Hong Kong JORC are all similar reporting formats that structure public disclosure. They are targeted at investors and do not specifically address project management of construction readiness.

SECTION 7

Viability Stage Implementation Opportunity to Pre-Feasibility or Feasibility

OBJECTIVE

This section defines the path for transitioning from the opportunity stage into the viability stage. It begins with a completed scoping evaluation developed into the formal document that meets *Project Management for Mining (PMfM)* standards and defines all activity required prior to the actual PFS or FS kickoff, including the development of battery limits and a complete set of final PFS or FS deliverables.

Each section of these guidelines is designed to direct the reader to the corresponding chapter in *PMfM*, which must be thoroughly understood for successful implementation. *PMfM* must be used alongside these guidelines.

REFERENCES

Item	*PMfM*	Page
Prefeasibility Study	Chapter 7 & 8	69
Feasibility Study	Chapter 8	83

CG Figure	Figure Caption	*PMfM* Reference
Figure 7.1	Project charter for scoping evaluation	New Figure

Cheklist TItle	*PMfM* Reference
Basic model considerations	New Checklist

SCOPE

Viability Stage Requirements

As a project advances from Scoping Evaluation to the viability stage (PFS or FS), a formal structure is established to determine the project's real contribution to corporate goals. That transition requires the following:

Scoping Evaluation

The Scoping Evaluation is complete and shows a probable reason for advancing to the viability study stage.

Management Decision for FS includes a 3rd party review.

A full management review is conducted to move to the viability stage with the goal of conducting a PFS or FS. Approval should be obtained to bring in an experienced project delivery expert to conduct a gap analysis based on the scoping evaluation results.

Conduct a Gap Assessment

A project gap analysis is conducted to evaluate all aspects of the project, including the inputs and outputs of the specific project. For example, if it is a process expansion or circuit addition, then the review should cover everything from the ore body to the final process output, and all related areas, to either confirm needed changes or confirm that no change is required.

Appoint a PFS or FS Leader

With the gap analysis completed, a competent and experienced project leader should be appointed (*CG* Section 4). This is a critical step to ensure all aspects of the project are managed correctly going forward.

Finalize the Strategy for the FS, Including Battery Limits

Finalize in detail, through a set of battery limits, who will be responsible for specific work. At this point, the division of work and scope must be very specific.

Develop PFS or FS Project Charter

- A comprehensive Project Charter with work plan is required going forward. This will clarify what work must be done by the operating entity prior to the PFS or FS kickoff, along with setting up the project correctly.
- Finalize all the details in an FS Project Charter document. This should include a definitive statement about the required accuracy of the FS, such as: "The FS estimate accuracy is to be +15% / –10%, with basic engineering well advanced beyond 85% and full engineering more than 35% completed." This statement must be included in all documentation related to hiring any contract support for the FS, including the lead engineering company.
- It is also critical that senior management understands and supports this level of accuracy to ensure a viable project estimate is produced.

Detail Components

With the battery limits defined, develop the structure around the PFS or FS program, including administrative procedures, contracting controls, and the RFP for the lead engineering company.

Develop Next-Steps Work Plan for the PFS or FS

Utilizing the gap analysis and the scope of work, divide the work into logical areas to determine who could complete the various activities. Determine the level of engineering and estimating that is required.

Level of Effort for Viability Stage

Incorporate the appropriate requirements to be completed for a PFS or FS as outlined in *PMfM* Figure 15.2 (*CG* Section 4).

Management Decision for PFS or FS Funding

With the FS Project Charter complete, prepare the AFE for approval of FS funding (*CG* Section 5).

Initiate Activity for PFS or FS

The PM initiates the activities to implement the FS, including assembling the team, gaining agreement with operating units on how to work with their personnel, and developing details around budget and schedule.

Appoint the Lead Engineer and Other Specialists

Prepare documents for hiring contracted entities (RFP and contract frameworks) and initiate the process of candidate identification, selection, and contracting.

Initiate FS Activities

Proceed to the kickoff and FS activity.

Data models required to achieve best practices results in viability stage

The goal is to generate data and build models that will become the basis of funding and go-forward decisions. The level of detail to which the model is advanced will be dictated by the study phase, the company's experience, and the maturity of the district.

The model roadmap may include:

- Mineral Resource Model
  - Interpreted geology layers with structure, lithology, alteration, and mineralization.
  - Complete an estimation of economic minerals, deleterious elements, geometallurgical markers, etc.
  - Ensure modeling complies with relevant public disclosure requirements and best practices.
  - Mineral resource evaluation should be done at the company's established resource pricing.
- Geotechnical Model
  - Use relevant geological interpretations, adjusted as required to meet geotechnical needs.
  - Include relevant rock strength and rock characteristics to accurately model the stability of the proposed mining method and inform proper design parameters.
  - The model may include relevant information for any post-mining comminution required for processing or recovery.
- Geometallurgy Model
  - Understand and include any elemental or categorical parameters that impact recovery, throughput, penalties, and other processing limits. This will provide planners with the resolution to optimize processing performance.
  - Characterize the geochemistry of product and waste streams for shipping considerations and environmental controls.

- Hydrology Model
  - Model or understand water supply, including groundwater, river diversion or capture, lakes, dams, etc.
  - Model or plan water management for mine dewatering, stability impacts, inflow, water balance, discharge, etc.
  - Models can be deterministic where risk or consequence is low, or where the level of study is low; or they can use stochastic modeling where risk or consequence is high and uncertainty is also high due to lack of data, inaccessible areas, etc.
- Mineral Reserve Model
  - Needs to combine the relevant modifying factors to correctly state the recoverable ore and inform the potential value of the project.
  - Mineral reserves should be estimated at the company's reserve pricing.
- Mine Planning Model
  - This combines reserve-model estimates with associated recoveries, mining dilution, and other modifying factors.
  - Modifying factors should be supported by other associated models with documented translations or scripts.
  - Costs and cutoffs should also be included for any routing requirements necessary in mine planning.
  - Any stockpiling strategy should be documented and coded into the model.
  - Capital, sustaining, and operating costs to achieve the production plan.
  - Economic models to determine cash flow and project value including the annual production of the projects economic outputs (metal, commercial materials, etc.).

Quality Control of Model Development

- Contracting for model development must include a due diligence review. It is probable that one will have to look outside the borders of the country to find qualified person(s) who can deliver a proven model and sign as a QP. For projects requiring external lending, consider using QPs with experience in the major project space who can provide appropriate reports to support the modeling.
- Understand that development of a model for the Owner is a large challenge. It must be understood that inexperienced modelers, or the unvalidated use of AI or implicit tools, could render the development model unreliable. That apparent unreliability may not be discovered until the project is up and running, which results in a failed project.

Common Problems for Projects with Mineral Reserve Estimates

- Sampling errors related to sample collection, sample mass, selected analytical method, or other QA/QC issues could render results unusable.
- Geological errors due to limited data, over-reliance on implicit modeling, lack of structural controls, etc.
- Modeling errors, such as geostatistics not being properly applied, may fail public or investor scrutiny and lead to project underperformance.
- Engineering errors in assumptions, process capacity or recovery, or mining rates could lead to aggressive or overly optimistic plans.
- Market evaluation errors, such as metal prices, penalties, or market requirements, could overvalue the project unrealistically.
- Gaps in the analysis process may affect what constitutes the full cost of the final deliverable.
- Improper modification of critical factors such as mineral recovery, ore dilution, benchmarked mining or processing costs, permit schedules, transport, in-country beneficiation rules, etc., can skew models so they do not accurately reflect the project's true value.

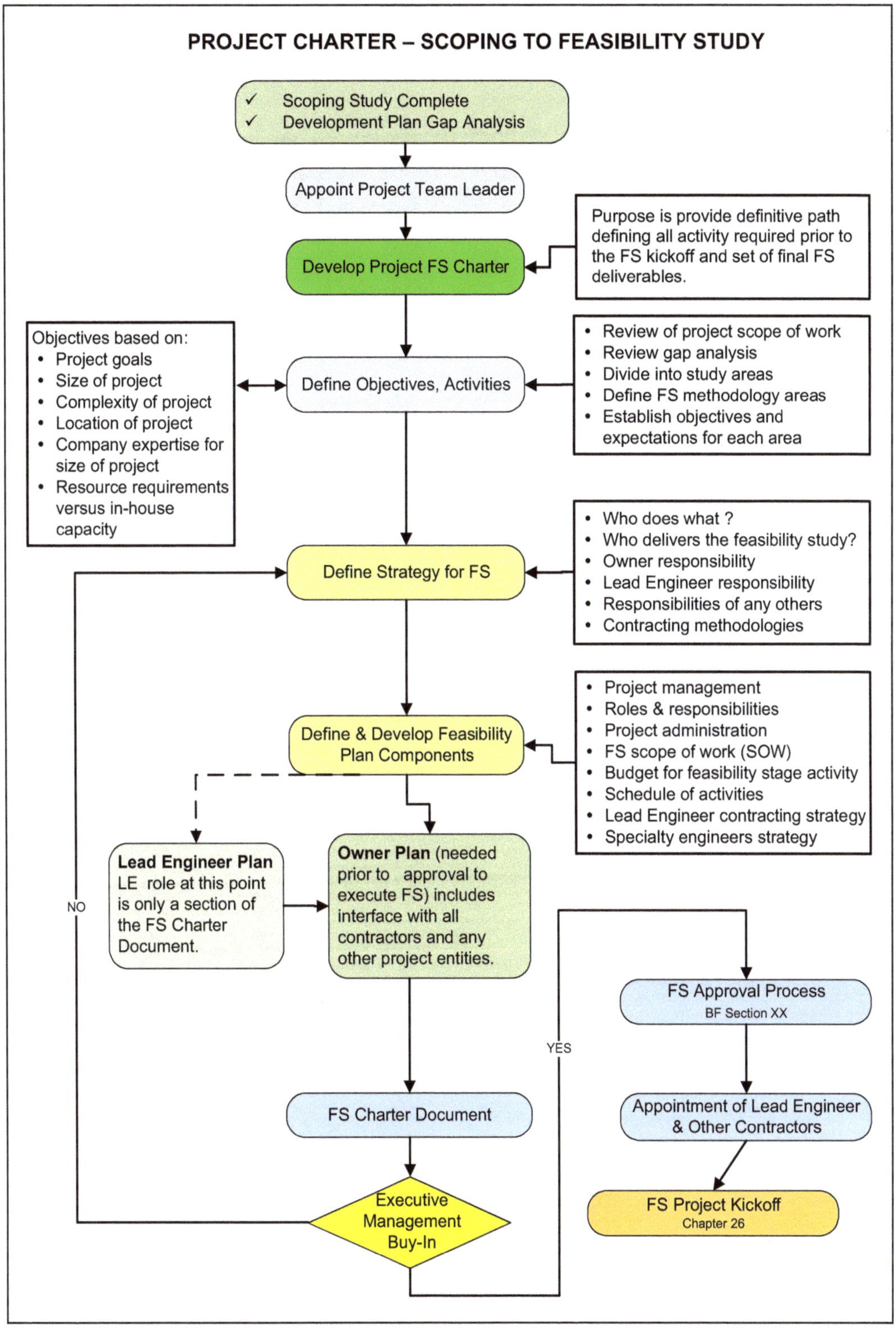

Figure 7.1: Project charter for scoping evaluation

BASIC MODEL CONSIDERATIONS

No	Item	Status	Date
	Database		
1	Data types (RC, DD, channel sampling, etc.)		
2	Data and test-work/assay quality (data quality management or QA/QC program, chain-of-custody, database controls, etc.)		
	Geological Modeling		
3	Structure modeling reflecting regional structures, any relevant mineralization controls, stability inputs, etc.		
4	Lithology modeling that respects data with >90% match to logging, 3D with section and plan alignment, and reconciles all available data.		
5	Alteration modeling relevant to mineralization controls or rock mass qualities.		
6	Mineralization domains based on structure-lithology-mineralization, geochemistry or mineralogy, or based on grade shells.		
	Resource Estimation		
7	Exploratory data analysis (EDA) to determine relationships between ore estimate and geological model		
8	Grade continuity and assumptions (variography matches geological interpretation and assumptions are valid for deposit type)		
9	Model checks and validation (industry best practices to ensure model construction is repeatable and fit for purpose for deposit type)		
10	Post-processing (dilution or other post simulation to better reflect recovered grade and tonnes)		
11	Resource classification (geometric, score-card, or other industry standard approach)		
12	Tonnes and grade reporting (should be consistent and ideally rely on block values to control outputs)		
	Geomechanical Modeling		
13	Relevant structures (bench or deposit scale, orientation, relative strength and properties)		
14	Rock strength (RMR/RQD and testwork)		
15	General mining method and geometry for stability analysis		
16	Geological inputs (lithology, contracts, alteration, etc.)		
17	Seismic study for additional design considerations		
	Geometallurgical Modeling		
18	Recovery (each economic element if possible and relevant mineralogical or geochemical controls)		
19	Deleterious element mapping (understand mineralogical groupings and distribution throughout deposit)		
20	Process water considerations (filtration requirements, chemical management, recirculating chemistry, etc.)		
21	Comminution energy (rock hardness and response to crushing, variability throughout deposit, impacts to processing throughput, etc.)		
22	Product and waste stream geochemistry (understand any shipping requirements and environmental controls)		

BASIC MODEL CONSIDERATIONS (continued)

No	Item	Status	Date
	Hydrology		
23	Inflow and outflow for modeling (sources and volumes)		
24	Hydraulic parameters for modeling		
25	Model extents (both directly accessible and potential area of influence)		
26	Users (communities, industries, and environmental)		
27	Water quality (filtration requirements, anticipated changes with abstraction or addition)		
28	Water balance for the site (water requirements, discharge, loss to environment, etc.)		
	Reserve Model		
29	Updated resource modeling using reserve pricing and costs		
30	Modifying factors aligned to the latest study phase		
	Mining Model		
31	Model to include mining dilution and other relevant modifying factors aligned with latest study phase		
32	Cost and pricing aligned with latest study phase		
33	Cutoffs and routing destinations based on modifying factors, costs, pricing, and any stockpiling strategy		
34	Stockpiling model or inventory tracking with spatial designs to ensure layout is compatible with other project infrastructure		
35	Mining extraction and material movement models for fleet sizing		
36	Mining development rates or sink rates documented and applied correctly to mining sequence		

SECTION 8

Feasibility Study Program

OBJECTIVE

The objective of the Feasibility Study is to quantify, within a universally acceptable structure, the proposed project's real contribution to corporate objectives. The study provides solid evidence of the project's feasibility (or lack thereof). It compiles the data necessary for making a yes or no recommendation to the appropriate levels of management on whether to proceed into the detailed engineering and construction stages that comprise project execution. An accompanying objective is to reach that yes or no judgment as early as possible, with no more money spent than necessary to reach an informed decision.

Each section of these guidelines is designed to direct the reader to the corresponding chapter in *Project Mangement for Mining (PMfM)*, which must be thoroughly understood for successful implementation. *PMfM* must be used alongside these guidelines.

REFERENCES

Item	*PMfM*	Page
Prefeasibility Study	Chapter 7	69
Feasibility Study	Chapter 8	83
Risk Management	Chapter 10	141
Scope and Work Breakdown Structure	Chapter 11	165
Schedules	Chapter 12	177
Budget	Chapter 13	201
Contingency	Chapter 14	229
Project Stage Work Efforts	Chapter 15	243

CG Figure	Figure Caption	*PMfM* Reference
Figure 8.1	Feasibility study	Flowchart 8.1, p.111

Checklist Title	*PMfM* Reference
Feasibility study	Checklist 8.1, p.108

SCOPE

Feasibility Study

A project's success can be directly measured by the effort put into the viability stage: the PFS and FS. Figure 8.1 identifies all the major elements that must be addressed during the FS process. It is important, up front, to have the battery limits defined, FS deliverables identified, and the expected level of accuracy in the final report established.

> *NOTE: Only by knowing the* required *FS end results (FS* deliverables*) can the detailed activities be scheduled to achieve those end results.*

An FS development plan must have a confirmed scope; therefore, a PFS may be required if there are major tradeoffs to be evaluated before the FS can be developed and approved. A less desirable method is to use the first months of an FS to conduct tradeoff studies. Once the scope is fully defined, the FS can be executed.

Prior to starting FS activities, it is recommended that the reader conduct a full review of *PMfM* Chapters 7 and 8, and *Companion Guidebook (CG)* Section 7.

FS Work Plan Kickoff Requirements

An experienced PM needs to lead the effort of implementing the work plan developed prior to starting the FS.

Data Collection

All data must be available to effectively start the FS, including ore body, metallurgy, survey, etc.

Infrastructure

The team evaluates whether any new infrastructure is included in the FS.

Project Execution Plan (*CG* Section 4 and *PMfM* Chapter 18)

Approximately midway through the FS process, the full PEP must be developed.

Owner Input

If it is a brownfield site, the Owner must provide key input in these areas as part of FS and PEP development (see *CG* Section 18).

Risk Management Program (*CG* Section 9 and *PMfM* Chapter 10)

A fully detailed Risk Management Program must be conducted.

Project Schedule (*CG* Section 4 and *PMfM* Chapter 12)

A Level 3 Schedule must be developed that is fully loaded, including the primary critical path.

Capital and Operating Cost Development (*PMfM* Chapters 13 and 14)

This is a bottom -up estimate that includes significant current data from operations, along with materials and construction costs.

Economic Evaluation (*CG* section 9)

The economic evaluation must be conducted by a knowledgeable person, either inside or outside the company, to make sure all financially related information is incorporated, including all Total Project Cost elements.

FS Completion

The completed FS should contain all the deliverables that were defined up front. The PEP should be in final draft form prior to AFE submittal.

Next Steps

Depending on the corporate mandate, the next steps could include executing engineering only, with procurement planning to get a better estimate of the full cost; or full execution, with engineering, procurement, and construction being released. The next steps are submitted as part of the Execution AFE defined in *PMfM* Chapter 14.

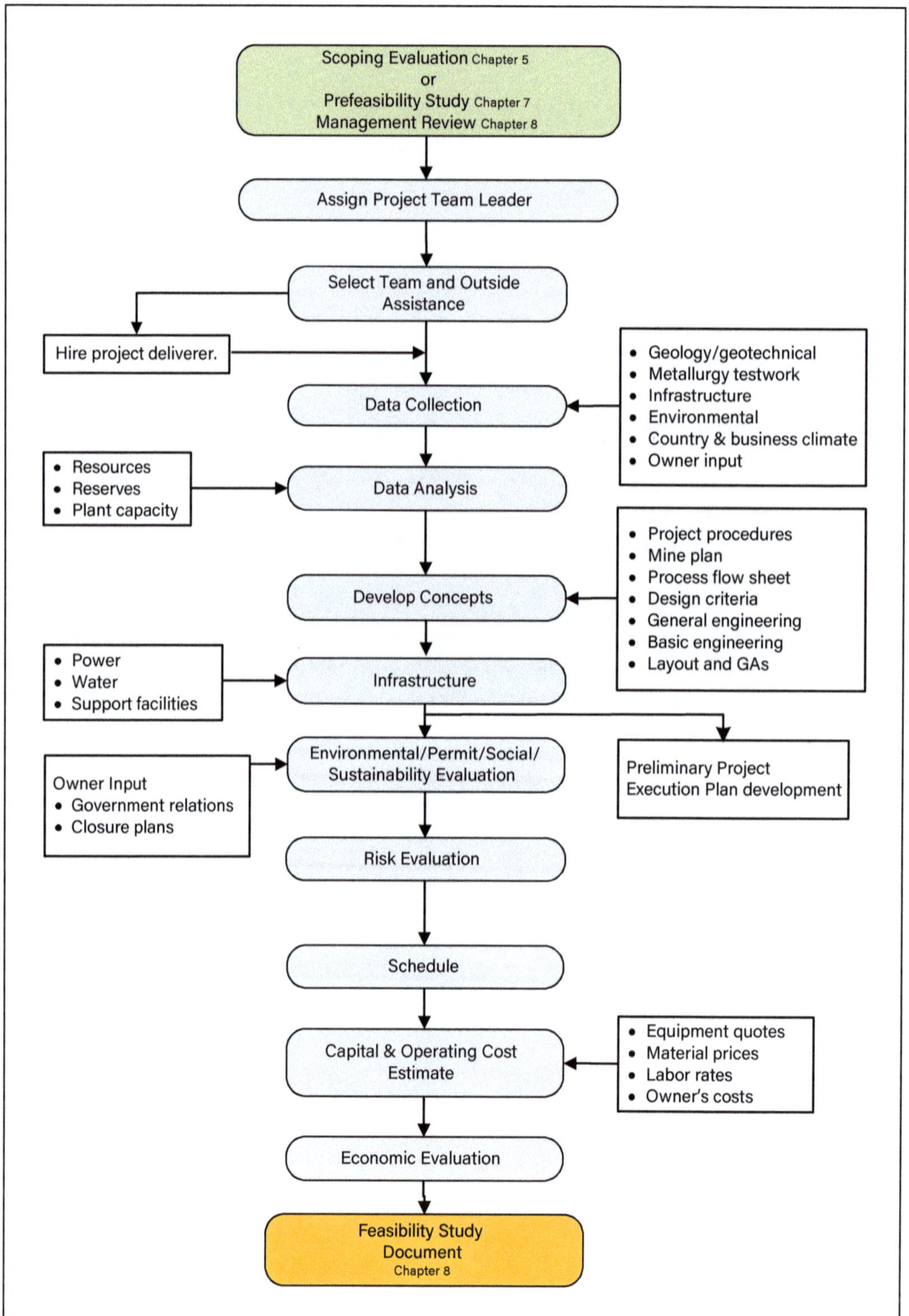

Figure 8.1: Feasibility study

FEASIBILITY STUDY

No.	Item	Status	Date
1	Completed scoping evaluation or prefeasibility study—in hand		
2	Management approval to proceed (per *PMfM* Chapter 6)—authorized		
3	Scope of work for feasibility study set and agreed to by Owner		
4	Project team identified and PM named		
	A. EPCM contractor support to project team—selected		
5	Property and site description		
	A. Data collection—undertaken		
	B. Background report—complete		
6	Mineral rights secured; land ownership/property lease—in place		
7	Field visits undertaken by:		
	A. Project leader		
	B. Core leadership group within project team		
	C. Key EPCM discipline heads		
	D. Owner's team (operations, business, and HR)		
	E. All project specialists (geotechnical, hydrology, etc.)		
	F. Power expert		
	G. Logistics and materials management		
8	Site constraints (location, climate, topography, access, etc.)—defined		
9	Geology program:		
	A. On-site mapping and sampling by geology team—done		
	B. Verification of existing data (including core assay checks)—OK		
	C. Confirmatory development and drilling—complete		
	D. Bulk sampling for metallurgical program—complete		
	E. Mineralogical study—undertaken		
	F. Geological model—created		
10	Mineral resource determined		
11	Mine reserve produced to CRIRSCO (e.g., NI 43-101/SEC S-K 1300) quality (2 derivations)		
	A. Cutoff grade defined		
12	Mining program:		
	A. Initial access to deposit—established		
	B. Geotechnical study—undertaken		
	C. Pit slope determination and stability study—complete		
	D. Pit prestrip and pioneering requirements—determined		
	E. Mining method—selected		
	F. Detailed mine plan—designed (including scope and criteria)		
	G. Mine equipment—selected		
	H. Mine services—detailed		
	I. Production schedule (annual and mine life)—complete		

FEASIBILITY STUDY (continued)

No.	Item	Status	Date
13	Metallurgical program—all requisite elements undertaken:		
	A. Metallurgical map of deposit		
	B. Ore characterization tests (hardness, abrasion, etc.)		
	C. Bench-scale lab tests on collected samples and/or new core		
	D. Process tests (float, grind, column, etc.)		
	E. Bulk (module) heap or leach tests—as appropriate		
	F. Pilot-scale plant test		
	G. Reagent selection tests		
	H. Salable product determination (composition and transport)		
14	Mineral process flow diagram—suitability confirmed		
15	Engineering design:		
	A. Soil and geotechnical reports—complete		
	B. Topographic maps (1:2,000 at 2-m contours)—in hand		
	C. Basic engineering—more than 85% complete		
	D. Value engineering trade-offs—complete		
	E. Overall engineering—more than 25% complete		
	F. Process flow diagram—determined		
	G. Mass and water balances—calculated		
	H. Design standards, criteria, and specifications—developed		
	I. Site facilities location layout map—established		
	J. Detailed GAs—produced		
	K. Equipment sizes and specifications—detail list complete		
	L. Process description—detailed narrative produced		
16	Infrastructure, support services, and utilities—needs identified and sized		
17	Hydrology—initial study completed		
	A. Dewatering strategy—in place		
	B. Water supply (process and potable)—determined		
18	Communications—system designed		
19	Power availability—initial study completed		
20	Access and transport (roads, rail, port, air)—initial study undertaken		
21	Project security—requirements documented		
22	Materials management:		
	A. Procurement plan—prepared		
	B. All long lead-time items—identified		
	C. Preliminary transport and logistics plan—undertaken		
	D. Warehousing and laydown sites—identified		
23	Workforce requirements—listed by job function and pay rate		
	A. Health and safety requirements—documented		
24	Training program needs—outlined		
25	Environmental and social impacts appraisal and quantification—done		

FEASIBILITY STUDY (continued)

No.	Item	Status	Date
26	Environmental studies:		
	A. Environmental specialist—on-site		
	B. Baseline data gathering—undertaken		
	C. Red-flag issues—identified and evaluated		
	D. Environmental monitoring and control program—initiated		
	E. Sediment and erosion plan—prepared		
	F. Reclamation plan—complete		
	G. Acid rock drainage study—complete		
	H. Geotechnical analysis for tailings and dumps—undertaken		
	I. Chemical reagents handling plan—produced		
	J. Characterization of environmental risks—undertaken		
	K. Environmental impacts and mitigation report—complete		
	L. ESIA/EIS/EA (or equivalent)—submitted		
27	Permits:		
	A. Existing and proposed regulations—fully determined		
	B. Approval schedule list—complete		
	C. Permit application submittals—initiated		
28	Social acceptance baseline study and impact analysis—complete		
	A. Social management plan—developed		
29	Sustainability and mine closure requirements assessment—produced		
30	Level 3 project coordination schedule (Gantt chart with work elements)—complete		
	A. Major milestones list—produced		
31	WBS—produced		
32	Project execution framework—outlined		
	A. Project controls system—outlined		
	B. Project reporting needs—defined		
	C. Quality management program—outlined		
33	Capital estimate (±15% accuracy at 90% confidence)—complete		
34	Operating cost estimate—complete, by area and by element		
35	Economic evaluation: IRR (pre- and post-tax) and NPV—finalized		
36	Market analysis—performed		
37	Currency exposure—defined		
38	Risk assessment (including country risk, if international)—complete		
39	Project opportunities analysis—complete		
40	Statement of project's competitive advantage—prepared		
41	Determination that project meets Owner objectives—done		
42	PDRI on project readiness—undertaken		
43	List of potential partners—prepared, if appropriate		
44	Financing options—identified, if required		
45	Feasibility study document—complete		

SECTION 9

Schedule, Budget and Contingency

OBJECTIVE

The project schedule provides the road map for completing the project successfully; it guides project accomplishment. The schedule is a dynamic tool that initially reflects the most realistic vision of the project's life span; later, it can respond effectively to changes in progress, scope, and resource availability over the life of the project.

The project budget details the cost estimate for the entire project scope through all project stages, from first study through viability, execution, start-up, and ramp-up to commercial production. A cost estimate is produced to approximate the cost of the project. It has a total figure with identifiable component values. The cost estimate is used to establish the total cost budget. The project budget is a necessity to support feasibility evaluations for project funding requirements and to facilitate project control. Total project cost versus initial capital is also explained in this section.

The correct application of contingency for a project is also explained in this section.

Each section of these guidelines is designed to direct the reader to the corresponding chapter in *Project Management for Mining (PMfM)*, which must be thoroughly understood for successful implementation. *PMfM* must be used alongside these guidelines.

REFERENCES

Item	*PMfM*	Page
Schedule	Chapter 12	177
Budget	Chapter 13	201
Contingency	Chapter 14	229
Authorization for Expenditure	Chapter 16	265
Executive and Board Decision	Chapter 17	271

CG Figure	Figure Caption	*PMfM* Reference
Figure 9.1	Schedule and cost comparison	New Figure
Figure 9.2	Estimation steps	Figure 13.1, p.202
Figure 9.3	Schedule	Flowchart 12.1, p.199
Figure 9.4	Budget	Flowchart 13.1, p.228

Checklist Title	*PMfM* Reference
Schedule	Checklist 12.1, p.198
Budget	Checklist 13.1, p.227

SCOPE

The project scope is the main driver for the project schedule and the associated development budget.

Scheduling

The project schedule describes what work is to be done, who will undertake the work, and when it should be done. How the work is to be done is defined by the PEP or Project Charter. Thus, the schedule does not control the work of a project; rather, it provides a guide for management to use in determining what focus to apply to the workflow.

Scheduling is a modeling process that can help communicate the chosen strategy and, at the same time, sequence and coordinate what is expected to happen in the future.

Projects should consider using 3D (three dimensional) modeling early in the design process. 3D models provide varying levels of detail in plan view as well as in all three viewing planes or axes. The three-axis nature of these models ensures that the spacing and relationship of elements are considered and provides a spatial reference to assist the project team in aligning on the layout. The ability to refine models and add additional dimensions is becoming standard in other project industries and is being brought into the mining industry.

- 4D adds construction time to help guide work fronts and timing for various stages of the project.
- 5D adds cost as a dimension. This helps confirm cash flows and allows stakeholders to analyze financial impacts along with design and build scheduling.
- Additional levels are being added to incorporate other layers into these models.

These 3D, 4D, and 5D elements can be incorporated into scenario modeling to understand bottlenecks and review build sequence alternatives to improve cash flow and/or build time.

Budget and Cost Estimate

Within the context of *PMfM*, the cost estimate is used as a forecast of the required funding for the project capital investment. The PEP, or Project Charter depending on the stage of the project, forms the basis for compiling a list of project tasks that are the activities required to complete the project work scope. The task list, which quickly transitions into the Work Breakdown Structure (WBS), has long been the foundation of project control because it is the basis that underpins the project schedule and project budget. Each task must have a definable work scope, a duration with at least one start date and one finish date, and a level-of-effort requirement to complete the task.

Both initial capital and working capital should be broken down into detailed cost allocations aligned to the level of study completed (i.e., scoping evaluation breakdown will be less detailed than at FS). This breakdown is key to reviewing costs, running cost sensitivity based on level of study, and aligning with existing company standards, particularly for brownfield projects. These cost estimates are only valid for the scope of work on which they are based.

This section details the steps needed to produce a project control budget as part of the FS. The corroboration efforts required to create *accurate cost estimates* for each study phase are outlined in *PMfM* Chapter 13, but the relevant details can be found in the individual chapters of *PMfM* that separately address each study. These study estimates are steps on the path to creating the project budget, and effort levels increase as the project progresses in complexity from Scoping, through PFS and FS, and into execution.

Definition Drivers

In preparing any capital cost estimate, the reality is that not everything will be known, specified, measured, or quantified. To prepare an accurate estimate, one must first assess what is known about the project and what is not known. The prepared costs will then fall into one of three categories:

1. What we know we know (the known knowns)
2. What we know that we do not know (the known unknowns)
3. What we do not know that we do not know (the unknown unknowns)

A project budget should account for the elements shown in Figure 9.2.

Total Project Cost

To correctly determine project contingency, management reserve, and funding requirements, it is necessary to ensure that all the components related to developing and starting up a new business are understood. Total Project Cost, as opposed to thse initial capital defined in a feasibility study is a critical distinction.

Total Project Cost provides all funding from the end of feasibility study when a project moves into and through the execution stage, including the commissioning stage, into start-up, and finally to achieving commercial production (i.e., reaching a cost neutral basis for a minimum of one calendar month).

The Total Project Cost estimate must consider and clearly identify the following list of typical cost components. These costs are itemized and included in the project economic analysis.

- Mine production and processing plant plan, including appropriate ramp-up with two critical milestones: (1) ramp-up reaching commercial production, which is generally in the throughput range of 70%+, but is set to take maximum advantage of the controlling tax laws; and (2) production throughput and output reaching 100%. (Following the start of commercial production, mine and plant expenditures become either operating expenses or sustaining capital.)
- Initial capital is defined in the FS as the costs required to build the facilities; both direct and indirect costs, including Owner's costs as defined in the PEP.
- Business and Operations readiness costs include all necessary components needed to operate an independent mine operation: manpower, training, systems, tools, etc.

- Any preproduction ore that will reside in inventory for processing once the plant is operating as part of "in process" inventory.
- Working capital, including costs for items such as raw materials, stockpile and in-pit inventories; warehouse inventory; insurance spares; delayed sales payments; and delayed accounts payable. These are costs that will be incurred prior to the start of commercial operations.
  - Warehouse inventory must include all the materials that the new operation will require, with appropriate inventory levels based on expected consumption of consumables, ordering timelines, receiving timelines, etc.
  - Routine maintenance and operational spares needed for the initial months of operation must be accounted for in the warehouse inventory build-up. Normally this is a 12-month supply, but it can vary from 6 to 24 months.
  - Long lead-time "insurance spares" crucial to operational start-up, but separate from routine maintenance spares, must also be considered. The monies allocated to this line item will vary substantially depending on the location of the facility. Suppliers provide most of the data needed to estimate this cost (it is critical to consider proven technologies to manage this cost).
- Capitalized operating costs incurred prior to commercial operation are capitalized as part of the project cost. These are costs associated with initial production, such as labor, power, consumables, and third-party services not included elsewhere in the project cost.
- Capital spares are major spare parts and redundant equipment components purchased as capital items for operations start-up.
- Management Reserve: The Owner may elect to establish an additional reserve when management requires even greater certainty that the project budget will not be overrun, or where there are significant identified project-specific risks that are outside the approved project scope and quality. Such a management reserve fund is sometimes authorized on top of project contingency to cover Owner uncertainties relating to execution of the project. These uncertainties could include political or social risks in the country of the project. Management reserve can best be described as monies that *may* be spent if a risk event occurs, whereas contingency is money that *will* be spent over the life of the project. The management reserve is approved by the Board and held by the CEO. The PM must provide evidence to the CEO that events have occurred outside the scope of the project in order to release these funds.
- Sustaining capital for the life of the property, especially focused on the first five years, detailed to support any final build outs of the property.
- Financing costs are the amount of money the company pays to obtain the required funding and include interest and principal payments.
- Escalation for Project Capital calculation: The project timeline is normally a three-to-four-year timeline. The economic model must include the expected escalation impact on the initial project cost, which is then added to the project contingency.

Contingency

Project contingency is best defined as "an amount of money for goods and services which, at the current state of project definition, cannot be accurately quantified, but which history and experience show will be necessary to achieve the given project scope. It is expected to be spent during the project execution stage."

The contingency estimate should include an evaluation of all areas identified in the Total Project Cost discussed above. Potential impacts to schedule, social conditions, and country risks should be considered. Schedule contingency is a real contingency that must be translated into additional dollars for the contingency calculation. The risk management plan must also be considered in the context of contingency for defined risks and their cost.

Calculating the required contingency requires conducting a Contingency Risk Analysis workshop with an experienced team. The workshop evaluates the WBS areas first on a qualitative basis, then on a quantitative basis, followed by running this data through @RISK software (or a similar product) that performs a Monte Carlo simulation. The output will produce the required 90% confidence level that is recommended as the final contingency value.

Contingency includes general, schedule, and risk management areas and is based on a Monte Carlo simulation to reach a 90% confidence level. It can be agreed that budget contingency is a sum of money added to a capital cost estimate to cover uncertainties within the estimate. Contingency is therefore expected to be spent to accomplish the project scope as defined.

Calculating contingency based on the process above for an FS that followed the best practice of reaching at least 30% to 35% engineering completed will result in a contingency value in the 15% to 18% range. A challenge for companies complying with U.S. SEC S-K 1300 rules is the requirement to report no more than 10% contingency within the published FS. This is unrealistic, and the project could not be completed with that reduced contingency value. This forces the company to add additional funds to a management reserve that would be expected to be used over the course of the project. See *CG* Section 5, p.3 for related details on total project funding.

Budget And Schedule Accuracy Context

Figure 9.1 provides a comparison of the different methodologies for levels of accuracy of schedules and budgets. There are five levels, and the most critical is level three which is what the completed feasibility study best practices require.

	AA	BB	CC	DD	EE	FF	GG	HH
	Schedules		Budgets					
	Schedules	Schedule Level and Detail	Budgets DOE / AACE (EPCM firms)	Budgets ASPE IPA	Estimate Name & Purpose	% Engineering in support of estimate	Methodology	Accuracy
A	Level 1	**Executive Summary Schedule** (1-page,top down schedule, minimal detail)	Class 5	Level 1	Order of Magnitude (Scoping) - FEL 1	0% to 2%	Factoring / Historical / Judgement	+/- 40% (-50% to +100%)
B	Level 2	**Management Summary Schedule** (cutomarily created as bar chart with more detail than level 1; up to 2 pages with some 50 line items) - [Top down or roll up from Level 3]	Class 4	Level 2	Prefeasibility Study Conceptual Design) - FEL 2	5% to 12%	Parametric* model (real attributes) Prelim. Takeoffs / Unit rates / Some factoring	+/- 25% (-30% to +40%)
C	Level 3	**Project Coordination Schedule** (comprehensive integrated CPM overview of project; used as key control tool in FS & PEP)	Class 3	Level 3	Feasibility Study Budget - FEL 3	25% to 40%	Semi-detailed Mostly bottoms up estimate	+/- 15% (-20% to + 25%)
D	Level 4	**Execution or Project Working Level Schedule** (expansion of specific areass of the Level 3 - normally developed by contractors and/or subcontractors; displays all the required activities)	Class 2	Level 4	Control Budget for project (detailed) Also bid tenders	60% to 80%	Zero-base detail; all bottoms up, by spread (quantites x labor cost)	+/- 10% (-10% to +12%)
E	Level 5	**Detailed Short-Term Schedule** (day-to-day work tasks mapped out in detail within 1,2,3 or 4-week schedules; used by contractor supervisors based on look aheads from their Level4 schedule activtties)	Class 1	Level 5	Definitive (Final check and / or Lump Sum Bid)	>85%	All by takeoffs, firm vendor prices, actual labor rates. No gaps filled from cost books	+/-4% to 8%
Ref: from PMfM handbook		Ref: Pages 189-191, 195	Ref: page 205	Ref: Page 206	Ref: Fig 13.4 Page 208	Ref: Fig 13.5 Page 208	Ref: Page 205	Ref: Fig 13.5 Page 208

Reference to figures and page number for PMfM 2nd Edition handbook

ASPE - American Society of Professional Estimators

AACE - Association for the Advancement of Cost Engineering International

IPA - Independent Project Analysis

DOE - US Department of Engergy

Parametric* estimating *model - is a mathematical representation of ***cost*** relationships that provide a logical and predictable correlation between the physical or functional characteristics of a project (plant, process system, etc.) and its resultant ***cost***.

Figure 9.1: Schedule and cost comparison

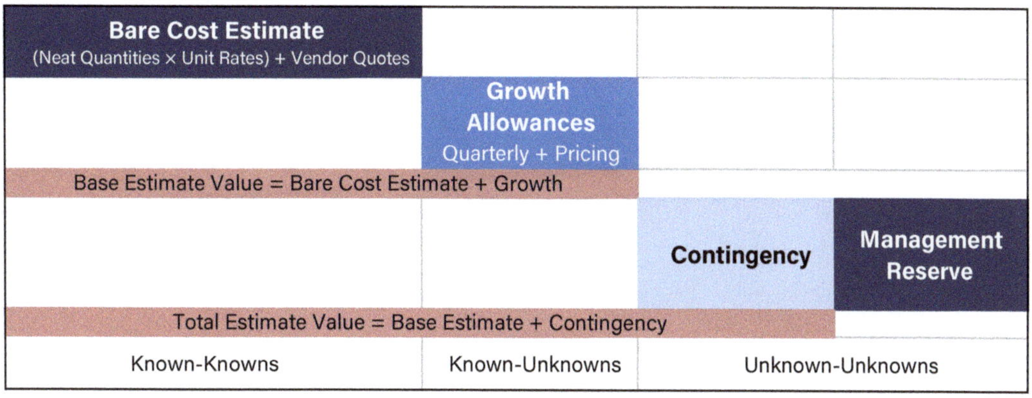

Figure 9.2: Estimation steps

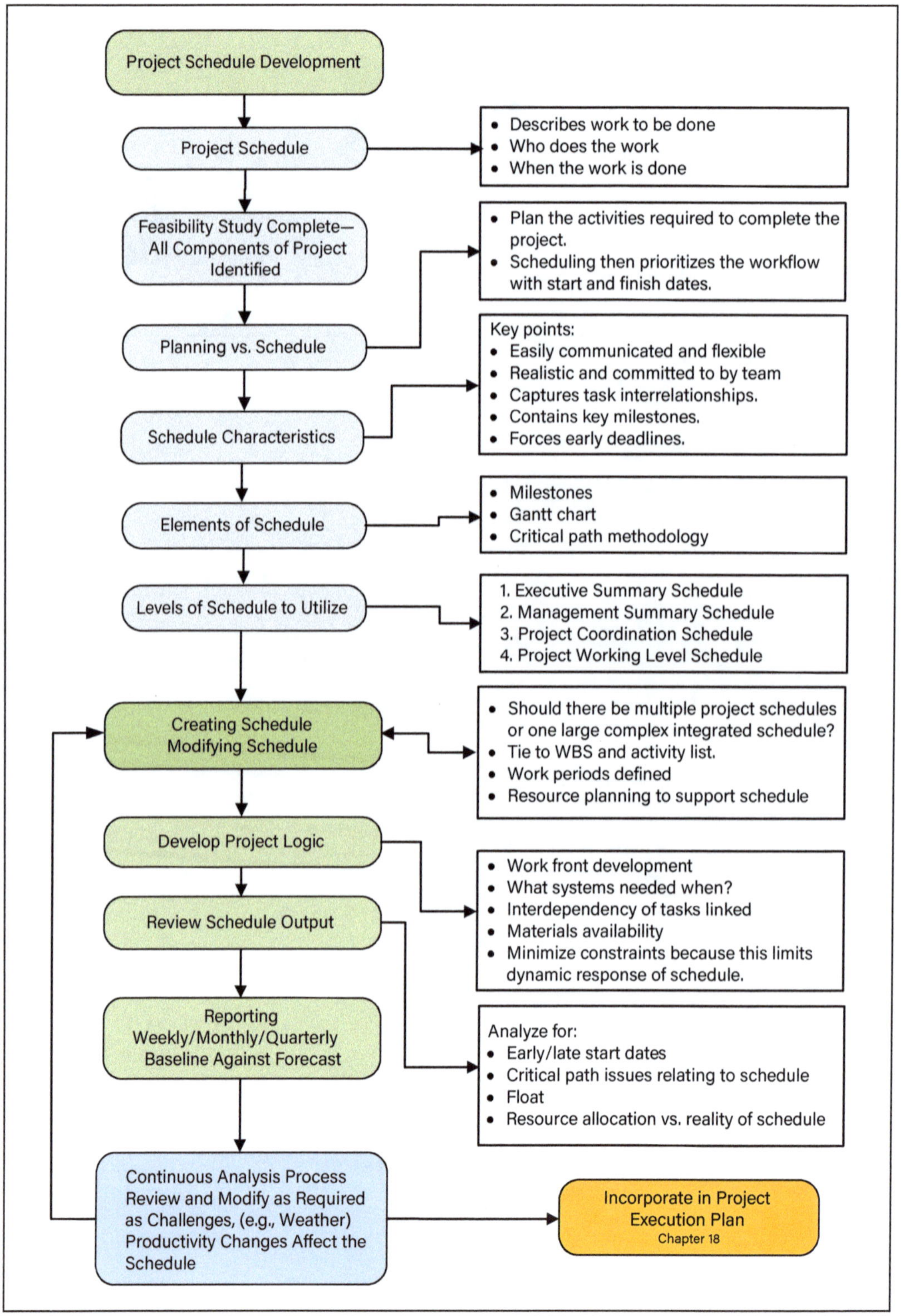

Figure 9.3: Schedule

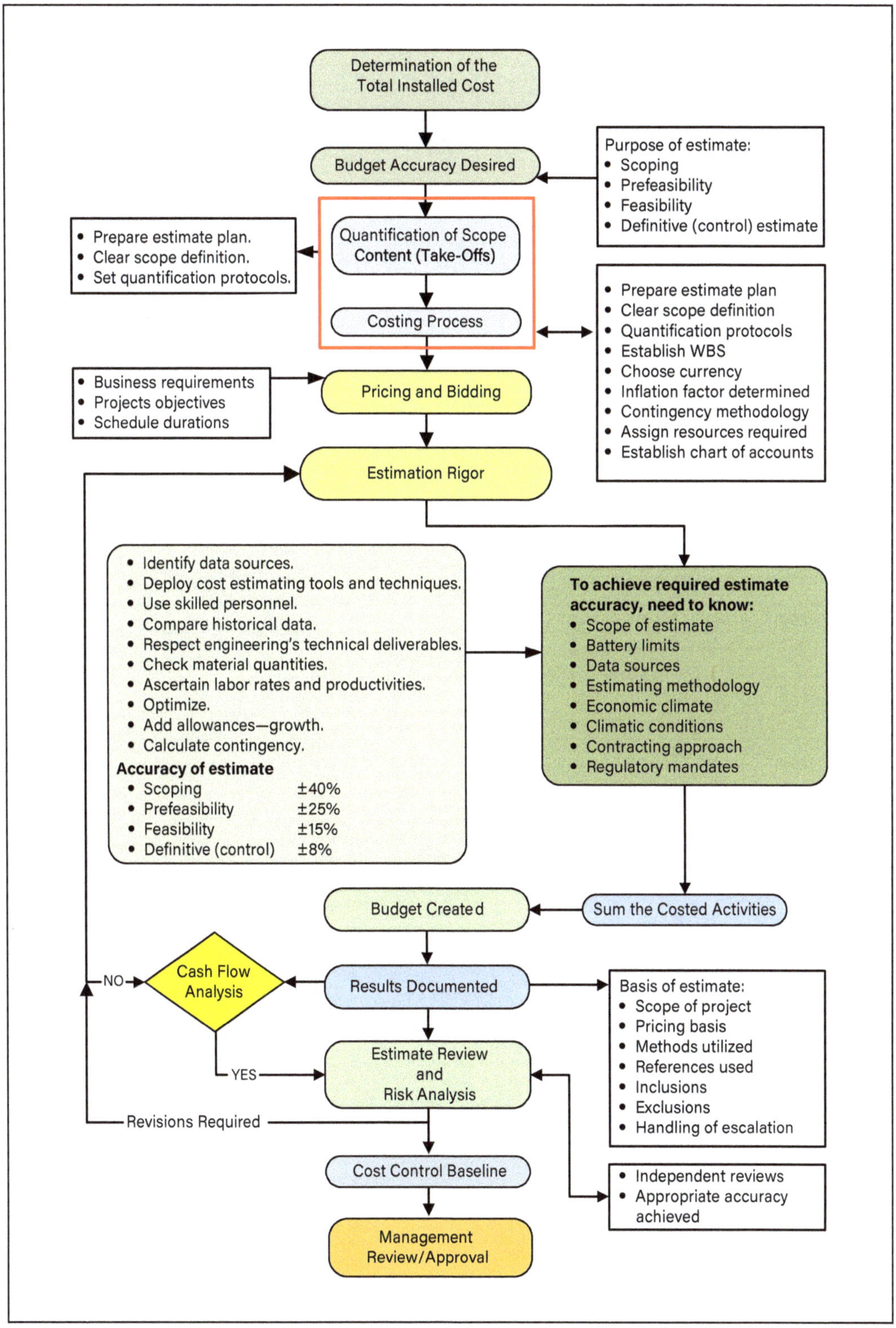

Figure 9.4: Budget

SCHEDULE

No.	Item	Status	Date
1	Level of schedule & detail required for stage of project—determined		
	A. Scoping—Level 1 executive summary		
	B. Prefeasibility—Level 2 management summary		
	C. Feasibility—Level 3 project coordination		
	D. Project delivery—Level 4 execution & level 5 detail		
2	Level of effort involved to achieve scope—understood		
3	Prerequisite work elements for schedule development:		
	A. Support study for the schedule—complete		
	B. Density of information required—understood		
	C. Scope of work—issued		
	D. Level of detail for schedule—defined		
	E. Level of accuracy of schedule—set		
	F. Planning of work process—completed		
	G. Task list—produced		
	H. WBS linkage—in place		
4	Schedule components		
	A. Project activiy list—complete (covers 100% of scope)		
	B. Project activity coding structure—prepared		
	C. Project calendar and work periods—set		
	D. Time scales for activity durations—selected		
5	Scheduling method or techniques selection:		
	A. Wall scheduling—utilized		
	B. Critical path methodology modeling technique—chosen		
6	Schedule development process:		
	A. Project logic—developed		
	B. Gantt chart—created		
	C. Task interrelationships—shown		
	D. Milestones—set		
	E. Free float—built in		
	F. Resource smoothing or leveling—undertaken		
	G. Schedule output analysis review—complete		
7	Schedule integration:		
	A. With cost management system—accomplished		
	B. With earned value management system—accomplished		
	C. With risk management system—accomplished		
8	Final schedule—peer reviewed		
9	Schedule update cycle—established		
10	Reporting requirements against baseline schedule—defined		
11	Schedule acceptance by owner and key stakeholders—achieved		

BUDGET

No.	Item	Status	Date
1	Purpose and classification of estimate—determined		
	A. Scoping		
	B. Prefeasibility		
	C. Feasibility		
	D. Definitive		
2	Level of effort involved to achieve scope—understood		
3	Level of accuracy required for the estimate type—quantified		
	A. Maturity of input information—realistically assessed		
	B. Project definition—established		
	C. Estimate quality—set		
	D. Confidence level (e.g., 90%)—selected		
4	Prerequisite elements for estimate development:		
	A. Project scope—clearly defined		
	B. Battery limits—set		
	C. Valid data sources—available		
	D. Estimating methodology—Selected		
	E. Economic climate—known		
	F. Climatic conditions at site—documented		
	G. Contracting approach—determined		
	H. Regulatory mandates—identified		
	I. Work breakdown structure—in place		
	J. Tasks and activities—clearly identified		
	K. Currency or exchange rates—established with Owner		
	L. Owner's accounting protocols—understood		
5	Costing process:		
	A. Estimate work plan—prepared		
	B. Strict quantification protocols—in place and followed		
	C. Pricing and bidding process—defined and followed		
	D. First principles base estimate (known knowns)—prepared		
	E. Growth allowances (known unknowns)—applied		
	F. Contingency allowance (unknown unknowns)—calculated		
	G. Basis of estimate—produced		
	H. Estimation rigor—undertaken		
6	Draft estimate—summarized		
	A. Final estimate—peer reviewed		

BUDGET (continued)

No.	Item	Status	Date
7	Full budget—created		
	A. Budget cost elements—established with Owner		
	B. Working capital and financing costs—prepared		
	C. Escalation treatment—decided		
	D. Confirmation of requisite accuracy—determined		
8	Budget results—documented		
9	Risk analysis—prepared		
10	Cash flow analysis—undertaken		
11	Independent reviews—conducted		
12	Project baseline control budget—established		
13	Total project cost has been incorporated into final estimate and financial analysis		

SECTION 10

Environmental, Permits, Social, and Sustainability

OBJECTIVE

The Project environmental, social, and sustainability aspects need to be managed from first opportunity through the project stages and into production in order to earn, and maintain, acceptance and approval. It is necessary to meet both corporate standards and satisfy international best practices. This will require varying degrees of interaction and education of the communities and governments depending on the level of development in the region in which the project is based. A robust approach will ensure you can move the project forward.

Each section of these guidelines is designed to direct the reader to the corresponding chapter in *PMfM*, which must be thoroughly understood for successful implementation. *PMfM* must be used alongside these guidelines.

REFERENCES

Item	*PMfM*	Page
Environmental, Permits, Social, and Sustainability	Chapter 9	113

CG Figure	Figure Caption	*PMfM* Reference
Figure 10.1	Environmental, permits, social, and sustainability	Flowchart 9.1, p.139

Checklist Title	*PMfM* Reference
Environmental, permits, social, and sustainability	Checklist 9.1, p.137

SCOPE

Environmental, social and regulatory aspects including permits, standards, and other compliance related topics are often linked together. This is increasingly important as the knowledge and connectivity of people continues to increase around these projects. Each element needs to be linked to the project footprint and ready to address each stakeholder's concern. Both from a limit of the company's influence and evaluating how the company does, or tries, to influence the governments and communities to act in accordance with international best practices.

It is critical that a robust and qualified team is brought on to manage the Environmental, Social and Regulatory aspects. This team needs representation in upper management and should be involved in the planning and engineering process to ensure the scope is properly captured, communicated to the relevant stakeholders and that feedback from the stakeholders is brought back to the project team to action.

Environmental

Objective

Manage the Environmental and Sustainability Factors of the Project.

- Use FS data with prior Scoping and PFS to facilitate creation of the Project Description which defines what the project is and is not. It will be used to evaluate environmental impacts, and to identify the key data and issues that must be managed.
- Terms of Reference (TOR) document is created and approved by lead agency responsible for the EA, EIS or ESIA, as required by the local and national governments, as well as international standards and lending requirements if applicable.
- Develop EA, EIS or ESIA that both matches and complements the FS.
- Collect Baseline Data
  - Most permits require timely and routine data collection and reporting.
  - A comprehensive set of baseline data can protect the project from unfair criticism or litigation in the future. The collection will take over two years to gather (usually mandates monitoring two of each season so start early)
    - Capture biodiversity information including endangered species, rare species, or species of concern. This includes flora, fauna, and ecosystems.
    - Track the community needs, at risk groups and indigenous people that may be present.
    - Hydrology is critical. What are the community needs, what are the projects' needs, and how are water resources managed within the region?
- Review the impact of any other activities that happen on the proposed project site. Illegal or informal use of the land for mining, trade, livestock, or hunting should be identified and a plan to address the social and environmental impacts should be put in place. These impacts should be quantified as part of the baseline assessment to differentiate them from the projects impacts.

Social Matters

Objective

Manage the Project to earn Acceptance and Approval for the project.

- Keep in mind, every employee is an ambassador for the project.

Focus Areas

Support and develop projects through a focused Social, Environmental and Community Development Strategy. Software packages are available today to track stakeholder interactions (*PMfM*, p.115). Some key areas are:

- Access to clean drinking water is key to supporting a healthy local population and attracting people to the region to support the project.
- Access to medical support should address key issues within the communities as well as support recruitment. Health screening for employment may find issues that need long-term support to resolve (Hepatitis, Tuberculosis, Malaria, etc.). Attracting people to a potentially remote area to work without medical care is far more difficult.
- Primary education should be government or locally managed. Brick and mortar structures are not schools without properly trained teachers with the necessary support. NGOs, governments, and other organizations that focus on nationally aligned curriculum are needed to operate and sustain these facilities. Also consider other social and economic factors; are children getting enough to eat to pay attention in school, are all youth granted access to education, or will the company need to work with communities to enable this?
- Secondary Education including pre-university, trade schools and advanced education may benefit the project and provide new livelihoods for communities.
  - Focus on the needs of the project and long-term activities that will sustain the community after the project leaves. Having a long-term sustainable plan post project closure will ensure that the community develops a robust economy that can survive without the project.
  - Consider supporting college programs by sponsoring students and/or graduate programs where students can enter the workforce and experience the mining world. Graduate programs and internships may need to be in other operations rather than at a greenfield project. These graduates become ambassadors for the company and the future leaders.
- Business development could be sourcing locally, micro-loans to setup companies that can support the operations and partnering of local companies with larger national or international construction companies. These partnerships can bring local support to national and international suppliers providing both the project and the local communities with improved opportunities.

Key Stakeholder Management

- Identify stakeholders and address each concern raised.
- Develop common ground with each stakeholder while managing their expectations.
- It is beneficial to move swiftly; but to obtain speed the project or company must first gain trust.
- Build trust with outside stakeholders and with inside management.

Stakeholder Engagement Program

- Set up a program that maintains transparent communication on key issues throughout project life (*PMfM,* p.255). Programs should be focused on updating the stakeholders on progress and activities, soliciting feedback from the stakeholders, and determining ways to engage stakeholders with the process to foster ownership in the project.
- Ensure that any indigenous peoples are identified and engaged with directly. Their concerns over social and environmental topics need to be managed for a successful engagement plan to enable the project to move forward.
- Embrace and work the plan by developing and putting in place all the necessary policies.
- Most programs will need to conform with IFC and *Equator Principles*; the minimum set of performance standards required by international project financiers

Social Acceptance to Mine

The perception held by the community can have a significant impact on how the mine operates.

- This is not a legal term. So how do we comply? The answer is through continuous engagement.
- Social Acceptance vs. Social Approval
  1. Social acceptance is *"the disposition of the community to tolerate, agree or consent to the project"* (This is the lower level of social acceptance, and is what a project seeks.)
  2. Social approval is when project earns the higher level of social acceptance, i.e., *"favorable regard of, agreeing to, or being pleased with project".*
- Social acceptance may not always suffice, and informed consent may be required.
  - Informed Consent is a process where, before proceeding, the project has to get community permission based upon their clear understanding of relevant facts & consequences.

Regulatory

Objective

To establish a sustainability plan that addresses permitting, ESIA commitments, and social programs that align to corporate standards, local and national governmental requirements, and meets IFC and World Bank standards.

Set Permitting Strategy

Create a permitting roadmap upfront to define realistic timelines and determine realistic work effort required to secure necessary permits.

- List local, state, regional and national permits/requirements along with a key milestones timeline (*PMfM,* Figure 9.1, p.117).
- Plan and implement a cradle to grave (i.e., beyond closure) strategy.

Project Requirements

Outline the relevant monitoring and management plans that the project will require.

- Create or review the "Construction Environmental Impact Management Plan".
- Implement a "Project Environment Monitoring Plan".
- The project team needs to understand the Owner's priorities, and the commitments already made, along with cost and schedule impacts (e.g., hiring of locals, community suppliers, indigenous peoples' obligations).

Post ESIA—Key Construction Phase Focus Areas

Area of Focus	Actions or Plans
ESIA Management and Monitoring Plans	Establish the control documents
Stakeholder Mapping and Engagement Plan	Inform and manage feedback from stakeholders
Grievance Procedure	Establish before construction starts
Emergency Preparedness and Response Plan	In place before construction
Labor and Work Conditions	Include contractors and supply chain.
Resource Efficiency and Pollution Prevention	Enforce permit provisos
Community Health, Safety and Security	Gain support and align company and community goals
Land Acquisition Plan	Plan needs to provide buffer and manage population influx
Involuntary Resettlement	Align program to IFC guidelines
Indigenous People	Create forums specific to these populations to engage with the company and have an active role in the scope and practices related to community and environment
Biodiversity Conservation, Ecosystem Services & Sustainability	Build a plan with offsets or projects to manage potential impacts
Cultural Heritage Preservation	Utilize local museums and universities and include intangible cultural elements
Litigation	Be prepared for this contingency as a last effort

Financing Considerations

Financing institutions are often leading the Environmental and Social aspects of industrial development. They often share related objectives, but each bank and each region will have different focuses. The project team needs to be aware of each institute's requirements and objectives to ensure they are addressing these points early and robustly. Engagement with the target investing institutes early will ensure they are engaged and part of the process rather than on the sidelines critiquing the work.

- Issues that IFC, World Bank and ISO require to be addressed:
  - Sustainability Development and Social Risk Management
  - Environmental Risk Control / Pollution and TSF Failure Prevention
  - Resource Efficiency (deterrence of short term "high grading")
  - Labor and Working Conditions
  - Community Health, Safety and Security
  - Land Acquisition and Involuntary Resettlement
  - Indigenous Peoples and Cultural Heritage
  - Biodiversity Conservation

  Note: If the company is international or seeking international funding, it will be evaluated on how well it adheres to these principles. Where the company relies on national assets for transportation, processing, or labor hiring, it will be evaluated on how it influences, or attempts to influence, the same adherence to these principles. The project must be aware of this and work with the Stakeholder Engagement Program to work through these issues.

- While most banks have similar requirements, ensure that the finance team and project leaders are aware of the requirements of the specific financial institutes that will be approached to work with the project.

Sustainability in Design

Look for modern and future ready technologies in the mine and processing facilities without relying on novel or unproven systems and technologies.

- Consider, if proven advantageous or for future updates, advanced tailings facilities (water conservation, concurrent reclamation, and provide for capacity to further treat tails if potentially valuable elements can be recovered).
- Consider electric and autonomous technologies in mining that can reduce carbon emissions.
- Review mine and stockpile/waste facility designs that have reclamation in design. This can save cost and lower environmental and visual impact.
- Ensure facilities and technologies are future proof or can be adapted to meet advances in design.
- Ensure local participation:
  - Utilize local sourcing and employment.
  - Consider developing support beyond the direct need of the company. This will ensure sustainable and long-term economies are developed that will continue after the project is completed and the community has a future beyond the project and operational life.

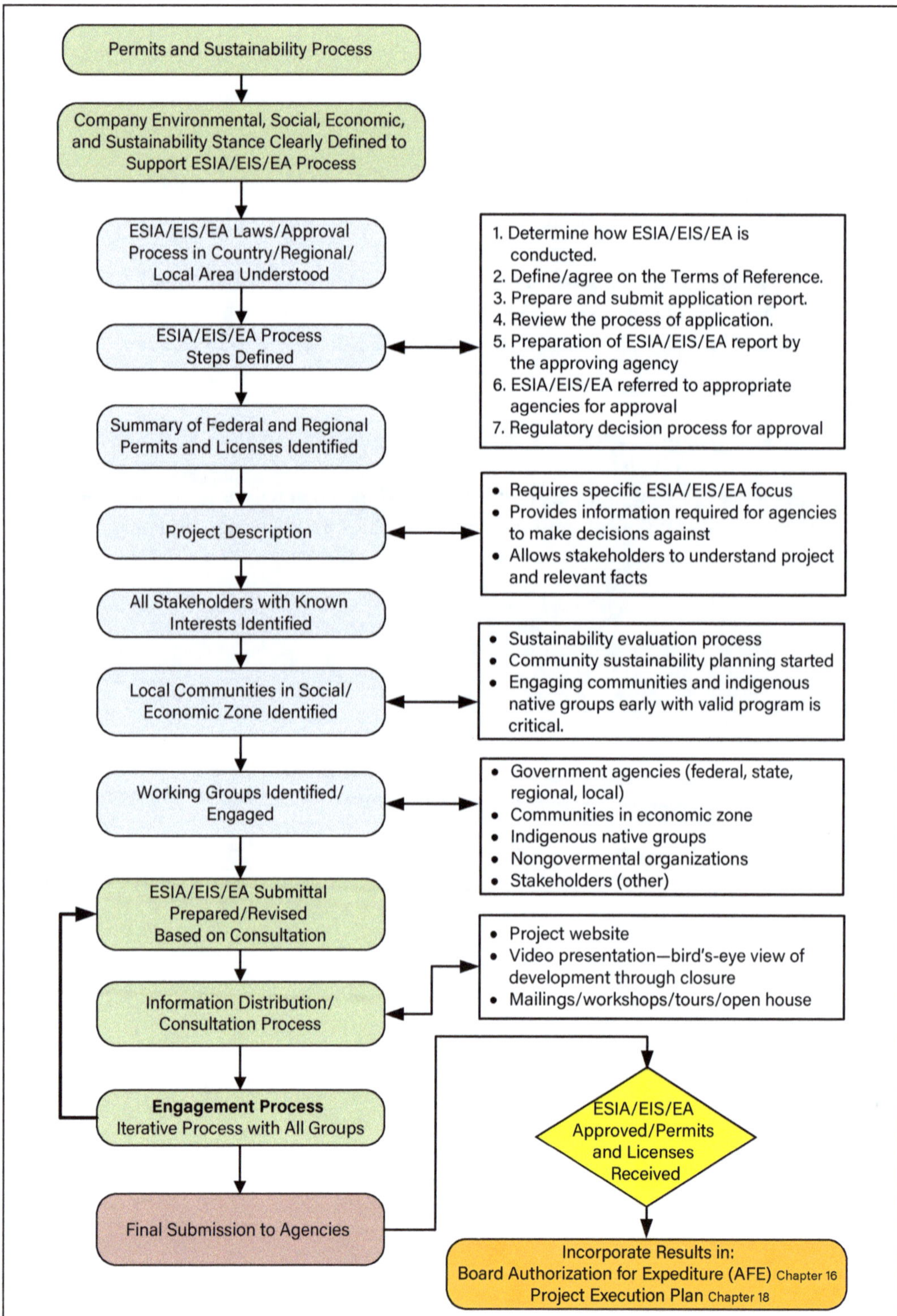

Figure 10.1: Environmental, permits, social and sustainability

ENVIRONMENTAL, PERMITS, SOCIAL AND SUSTAINABILITY

No.	Item	Status	Date
1	Owner's environmental, social & sustainability stance—understood		
	A. Environment, social & sustainability charter—in place		
2	Familiarity with current situation at the location of project:		
	A. Scoping evaluation, prefeasibility and feasibility studies—reviewed		
3	Permit regime that governs the project locale—fully comprehended		
	A. Federal (existing and proposed regulations)—determined		
	B. Regional (existing and proposed regulations)—determined		
	C. Local (existing and proposed regulations)—determined		
4	Permit or license list (federal, regional, local)—prepared		
	A. Permitting steps—clearly and realistically documented		
	B. Environmental and permit constraints/risks—identified		
	C. Permit schedule—established (with submittal dates)		
5	Environmental program:		
	A. Environmental specialist team—on site		
	B. Baseline data gathering—undertaken (minimum 2 years)		
	C. Environmental monitoring and control program—initiated		
6	Environmental studies:		
	A. Sediment and erosion plan—prepared		
	B. Reclamation plan—complete		
	C. Acid rock drainage study—complete		
	D. Geotechnical analysis for tailings & dumps—undertaken		
	E. Chemical reagents handling plan—produced		
	F. Characterization of environmental risks—undertaken		
	G. Environmental impacts and mitigation report—complete		
7	Key environmental management permit (EA, EIS or ESIA)—determined		
	A. Project description—completed		
	B. Terms of reference—accepted by regulatory agencies		
	C. Ancillary and support permits—identified		
	D. Project impact appraisal and quantification—undertaken		
	E. Red flag issues—identified and evaluated		
	F. Conformance to Equator Principles, IFC, and World Bank standards—achieved to the extended demanded		
	G. ESIA / EIS / EA (or Equivalent)—submitted		
8	Stakeholders (internal and external)—identified and contact established		
9	Local communities, indigenous peoples, and interested parties:		
	A. All community parties—cataloged		
	B. Project working group organization—in place		
	C. Community relations program—established		
	D. Critical external stakeholder engagement—achieved		

ENVIRONMENTAL, PERMITS, SOCIAL AND SUSTAINABILITY (continued)

No.	Item	Status	Date
10	Social acceptance issues—documented		
	A. Baseline study and impact analysis—complete		
	B. Potential negative social impacts—identified		
	C. Social value positive contributions—communicated		
	D. Consultation process—established		
	E. Media relations person—appointed		
	F. Stakeholder mapping and engagement plan—developed		
	G. Issues requiring acceptance by external stakeholders—identified		
	H. NGO support—garnered		
	I. Social management plan—developed		
11	Sustainability process—initiated		
	A. Sustainability review—complete		
	B. Sustainability management plan—produced		
12	Closure plan—produced		
	A. Likely environmental impacts—assessed		
	B. Active or passive care programs—developed		
	C. Implementation costs—estimated		

SECTION 11

Risk Management

OBJECTIVE

The objective here is for the project team to understand the risks of a mining project and to apprise the team of the process used to address the factors that underlie these risks.

A risk management framework needs to run through every stage, from the first scoping study through all five stages of the project, to ensure that long-term decisions, as well as day-to-day operational choices, are all made objectively with the project's goals in mind. *Project Management in Mining (PMfM)* Chapter 10 discusses this in detail.

Each section of these guidelines is designed to direct the reader to the corresponding chapter in *PMfM*, which must be thoroughly understood for successful implementation. *PMfM* must be used alongside these guidelines.

REFERENCES

Item	*PMfM*	Page
Risk Management	Chapter 10	141

CG Figure	Figure Caption	*PMfM* Reference
Figure 11.1	Risk management	Flowchart 10.1, p.164

Checklist Title	*PMfM* Reference
Risk management	Checklist 10.1, p.163

SCOPE

Risk Management Key Principles

Project Risk is an unforeseen event or activity that can affect project progress or outcome in a negative way.

Risk management identifies, quantifies, and manages those risks that can prevent an organization from reaching its objectives.

- Risk Management involves the handling of unacceptable risk.
- If risk is acceptable, it does not need active management.

Managing risks starts with creating a Risk Management Plan as part of the PEP.

- The purpose of the Risk Management Plan is to sort risks into action groups.
  - Those needing monitoring only.
  - Those requiring proactive management.
  - Those best mitigated by "transfer" (e.g., insurance) to other parties.

Risk Management Overview

The Risk Management Program deals with project risk by methodically undertaking the following steps:

1. Risk identification, which requires creation of a risk register.
2. Risk prioritization of the groupings of risks into logical categories.
3. Risk quantification for scoring likelihood and potential impact.
4. Risk analysis to assess "what can go wrong" using reasonable assumptions.
5. Risk mitigation, in which the chosen approach for risk treatment is implemented in four basic ways:
   a. Avoidance — Take action to avoid the risk from occurring.
   b. Alleviation — Carry out defined actions if and when the risk occurs.
   c. Transfer — Have someone else handle the risk by taking on additional insurance or bonding.
   d. Acceptance — Identify the risk as acceptable, then allow it to occur while monitoring it.
6. Risk monitoring for tracking risk mitigation activities.
7. Risk reporting for status reviews and Risk Register updates on a recurring basis.
8. Risk ownership assignment of accountability for developing and implementing mitigation plans.
   - Risks are assigned to the entity with the best chance of minimizing that risk.
   - The Owner still pays for risk management. The Owner cannot shed the cost of risk onto others.

Tasks or Role of PM in Risk Management

- Establish the Risk Management Plan very early in project life.
- Project, operations, and corporate leadership jointly undertake risk prioritization, quantification, analysis, and ownership assignment.
  - Determine any risk interdependency. Are there unavoidable cascading effects from these risks?
  - Evaluate the options and then calculate potential costs for selected mitigation actions.
- The PM is fully responsible for mitigation, monitoring, and reporting.
  - Establish key performance indicators for monitoring the highest risks.
  - Focus on the highest risks and understand the Owner's true risk tolerance threshold.
- With guidance from the corporate risk group, determine which project risks are best addressed by insurance coverage.
  - Typically, the major economic risks are mitigated, at least in part, by insurance.
  - The project cannot be, nor is it appropriate to be, insured for everything on the project site.
  - The PM must understand the advantages and limits of the various insurance packages.
- Have all insurance in place prior to the project execution stage.

Likely Project Risks

- Scope Risk includes mischaracterizations, gaps, and scope creep.
  - Risks may include unanticipated regulatory and political changes.
  - Risks may include unexpected project participant behavioral changes.
- Schedule Risk may include estimate errors and untimely decisions.
  - Risk may be related to equipment malfunctions or parts shipment delays.
  - Risk may be triggered by project dependency failure, where one or more parties are not delivering work on time.
- Resources Risk includes insufficient funds for items or staffing.
  - Late workforce or contractor buildup, along with skills scarcity and training shortfall, can impact project delivery.
  - Resource attrition through personnel turnover or desertion could lead to productivity deficits.
  - Risks could come from corruption and fraud issues or differing joint venture partner agendas.
- Social Acceptance to Operate could be impacted by community activism or security issues (real or perceived).
- Competing Demands regarding land, access to the project site, water resources, energy resources, and contractor availability can all impact project delivery.
- Capital Access may change based on commodity price, currency volatility, or inflation.
- Resource Nationalization may threaten a project, particularly if there is a real or perceived discrepancy in benefits sharing.

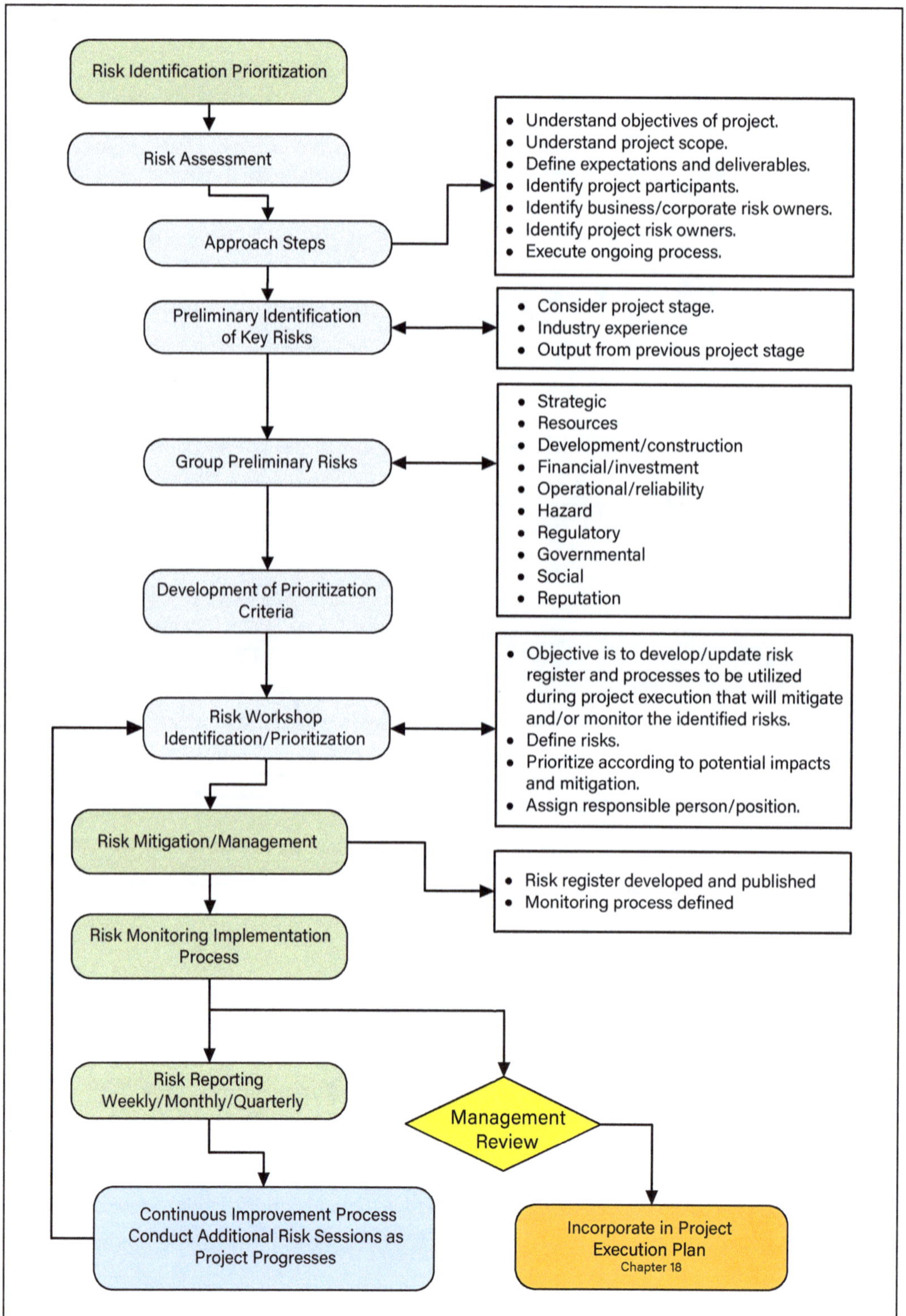

Figure 11.1; Risk management

RISK MANAGEMENT

No.	Item	Status	Date
1	Project objectives, expectations, and deliverables—understood		
2	Risk identification—complete		
3	Project risks addressed by a typical project:		
	A. Ownership		
	B. Installation cost (including inflation concerns)		
	C. Constructability		
	D. Infrastructure and utilities access		
	E. Operational		
	F. Labor (including skills availability)		
	G. Geographic (including climate issues)		
	H. Geological		
	I. Technological		
	J. Environmental impact, permits, social acceptance, and sustainability		
	K. Foreign currency exposure		
	L. Political, country, and nationalization business practices		
	M. Tax and business climate		
	N. Economic, financeability, and capital access		
	O. Project execution		
4	Risk register—published		
5	Risk categorization and prioritization—completed		
	A. Risk workshop—conducted		
6	Risk criteria quantification (likelihood and potential impact)—scored		
7	Risk analysis—undertaken		
8	Risk matrix—produced		
9	Risk ownership accountability—assigned		
10	Risk mitigation or treatment—action plan developed		
	A. Proprietary software—selected		
	B. Implementations initiated (either avoidance, alleviation, transfer, or acceptance)		
11	Risk monitoring		
	A. Key performance indicators—developed		
	B. Monitoring individuals—selected		
12	Risk reporting—initiated		
	A. Reporting format—established		
	B. Status update frequency—determined		
	C. Risk dashboard—created		
13	Bow-tie method of risk evaluation—used for those potential hazards that can have a significant effect on project outcome		
14	Owner's risk tolerance threshold—established and known by project manager		

SECTION 12

Project Execution Plan

OBJECTIVE

The Project Execution Plan (PEP) is the document that outlines how the project will be executed from full project approval through completion and turnover. Development is undertaken during the FS activities and prior to the development of schedule and cost details.

The FS defines the technical aspects of the project, but it is the PEP that explains in detail how the project will be developed. The PEP is a required document that must be completed prior to preparation of the Execution AFE. The PEP is not a chapter within the FS, but a standalone document.

The Project Charter, developed after the scoping evaluation is a simplified precursor to the PEP because it deals only with execution of the activities around an FS.

Each section of these guidelines is designed to direct the reader to the corresponding chapter in *Project Management for Mining (PMfM)*, which must be thoroughly understood for successful implementation. *PMfM* must be used alongside these guidelines.

REFERENCES

Item	*PMfM*	Page
Project Execution Plan	Chapter 18	277

CG Figure	Figure Caption	*PMfM* Reference
Figure 12.1	Project execution plan	Flowchart 18.1, p.298

Checklist Title	*PMfM* Reference
Project Execution Plan	Checklist 18.1, p.296

SCOPE

As the FS progresses, the full scope of the project becomes evident, allowing all related areas to be modified and the full project scope to be identified. The objectives for the project can also be clearly defined. These form the basis for development of the PEP (Figure 12.1).

In a brownfield setting, it is recommended to hold a planning session with both the project team and operations personnel to ensure that all areas of the project, and all interfaces with operations, are clearly identified for the PEP.

Define the Execution Strategy

- The brownfield environment allows for a variety of methods to execute the project. It can range from a full EPCM model to an Owner PM team managing contractors and supported by existing company personnel. A careful review of the options in *PMfM* Chapter 19 is recommended.
- Contracting is one of the complexities of mining projects, with varying amounts of contracting undertaken depending on the size and scope of the project. An excellent way of communicating a complex project-delivery / contracting strategy is through the use of a "horse blanket," which is a cross-functional chart such as that displayed in *PMfM* Figure 31.1, p.574. Whenever the project work scope is segmented into multiple packages for different contractors to execute, it is vital that the segmentation be clearly conveyed to all project participants so that each participant understands their role in relation to the other parties. This is even more important whenever a hybrid project-delivery strategy has been chosen, such as one that utilizes the Owner, an EPCM, and perhaps an original equipment manufacturer (OEM) to deliver different parts of the project. *Note: Gap analysis by experienced project personnel is critical to ensure that each package* aligns *and* forms *the whole when complete. The more packages* there are, *the more potential gaps there* will be.

Developing the Execution Plan Components

This is a significant work effort and demands clarity of purpose for each section of the PEP.

- Differentiate between the Owner Execution Plan and the Contractor Execution Plan. With a known execution strategy, an initial development concept (prior to hiring contractors) must be assembled by knowledgeable project and construction personnel.
- The draft PEP is assembled and reviewed by the project team and operations related personnel to ensure all aspects of the project are covered. This detail is then used during schedule and cost development activities in the FS process.
- The PEP should be reviewed to ensure the project is ready for construction. The Construction Industry Institute has developed a *Construction Readiness Assessment (CRA)* that can be used. A CRA score greater than 85% is necessary for the project to be considered "ready for construction." Any deficient areas need to be addressed before project execution.
- It is imperative that executive management agrees with the principal strategy and objectives of the PEP.
- This document becomes a key part of the AFE approval process.

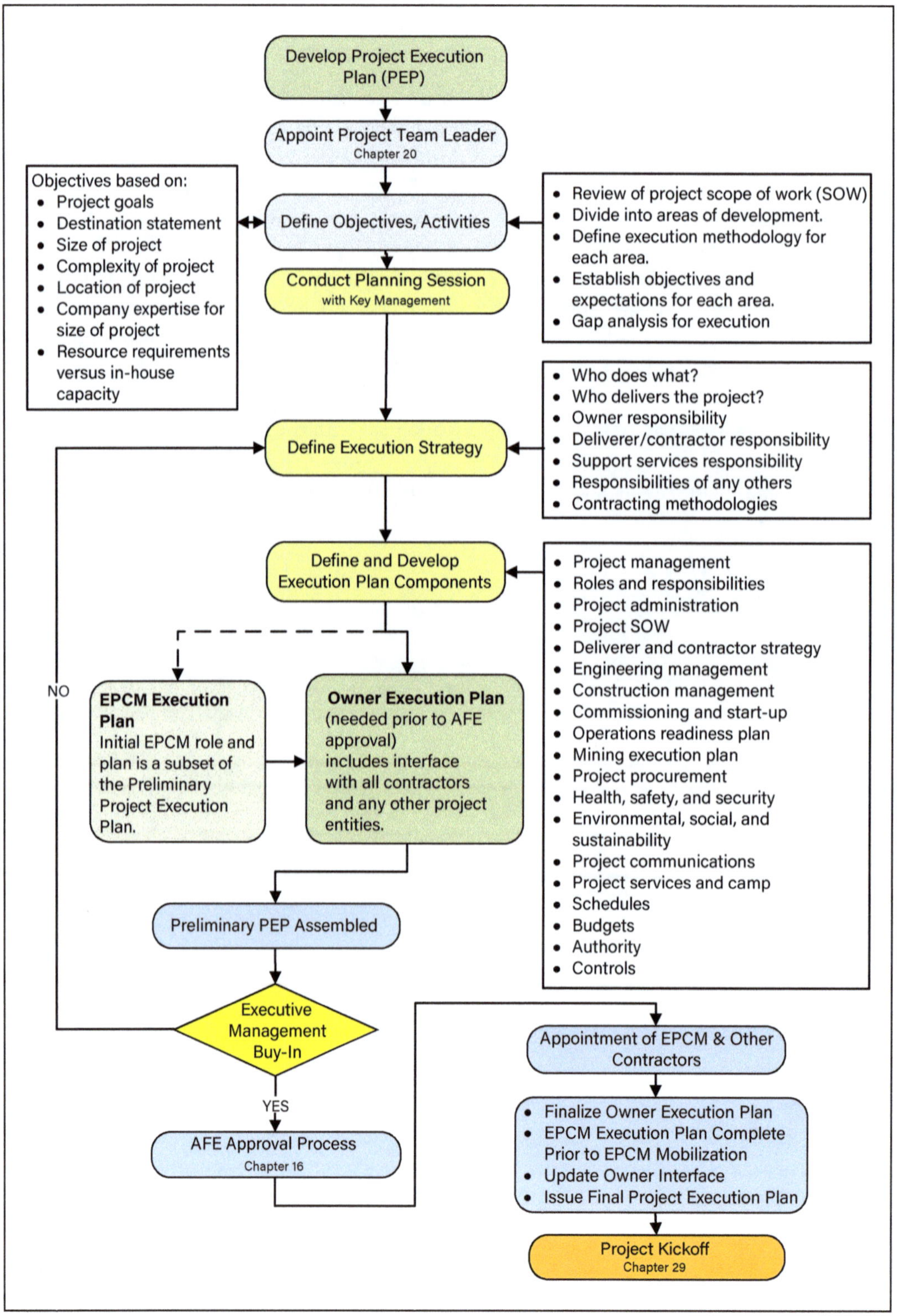

Figure 12.1: Project execution plan

PROJECT EXECUTION PLAN

No.	Item	Status	Date
1	Feasibility study complete		
2	Success criteria and requisite deliverables—established with Owner		
3	Project delivery method (EPC, EPCM, OEM, and/or Owner)—determined		
4	PEP—updated and expanded to final		
	A. Project description (synopsis of prior work)—documented		
	B. Definitive scope of work going forward—set		
	C. Key project goals—identified		
	D. Project criteria, battery limits, and constraints—identified		
	E. Gap analysis—complete		
	F. Owner's execution plan and deliverers' execution plan(s) coordinated into a single PEP—complete		
	G. Interactive project planner session—undertaken		
	H. PPM—assembled		
	I. Deliverables verification plan—prepared		
	J. Owner's executive management buy-in—obtained		
5	Project governance—established		
	A. Levels of authority documentation—in place		
6	Team selection strategy (internal and external)—set		
	A. Project leader—appointed		
	B. Core project management team personnel—identified		
	C. Project team organization RASCI chart—produced		
	D. Owner's technical support resources—assigned		
	E. Labor force availability (skills assessment, and hiring strategy) outlined with Owner's HR department		
7	Quality management plan—QA/QC protocols outlined		
8	Risk management plan—in place		
9	Project master schedule—published		
	A. All key milestones—established		
	B. Critical path—determined		
10	Capital cost management—outlined		
	A. WBS—in place		
	B. Budget allocation and cost reporting—fully detailed and set		
11	Project controls—established		
	A. Control system, resources, metrics, and functions—in place		
	B. EVM progress and performance measurement—in place		
	C. Scope, cost, and schedule control processes—defined		
	D. Change management system—set		
	E. Contingency control process—outlined		

PROJECT EXECUTION PLAN (continued)

No.	Item	Status	Date
12	Major accounting controls—in place		
13	Project administrative procedures—established		
14	Reporting requirements—defined		
15	Communications plan—established		
16	Project reviews—identified		
	A. Project dashboard—functioning		
17	Engineering and design requirements—established		
18	Procurement strategy and procedures—established		
19	Construction management approach—issued		
	A. Contracts management plan—prepared		
	B. Contract types to be employed—selected		
	C. Project breakdown into areas of development—outlined		
	D. Support services—identified		
	E. Health and safety requirements—identified		
	F. Site security needs—outlined		
20	Operations readiness plan—updated		
21	Environmental impact, social acceptance, sustainability, and closure plans—produced		
22	Commissioning and start-up plan—prepared		
	Operations workforce hiring and training program—defined		
23	Turnover to operations process—outlined		
24	Project financing plan – identified if appropriate		

SECTION 13

Controls

OBJECTIVE

Establishing sufficient controls early, appropriate to the stage of the project, is a key element of the overall project management process. This effort ensures the project is correctly defined, estimated, scheduled, and resourced. Project control requires constant use of short-term data and controls (daily and weekly) rather than belated reliance on monthly information.

Two critical points:

1. Project Controls are not Accounting Controls. See *PMfM* Chapter 24.
2. Project Controls differ from Operations Cost Management, and modifications may be required depending on the project stage to adequately control the project.

Implementation of adequate controls before the project gets underway will generally mean the difference between project success and project failure.

Each section of these guidelines is designed to direct the reader to the corresponding chapter in *PMfM*, which must be thoroughly understood for successful implementation. *PMfM* must be used alongside these guidelines.

REFERENCES

Item	*PMfM*	Page
Project Controls	Chapter 24	427
Reports and Progress Review	Chapter 25	457
Quality Assurance and Quality Control	Chapter 26	471
Document Management and Document Control	Chapter 27	487

CG Figure	Figure Caption	*PMfM* Reference
Figure 13.1	Project controls system	Figure 24.1, p.430
Figure 13.2	Project development and control process	Figure 24.3, p.436
Figure 13.3	Project controls	Flowchart 24.1, p.456

Checklist Title	*PMfM* Reference
Project controls	Checklist 24.1, p.454

SCOPE

Accounting vs Project Controls

To fully discuss controls for a project, it is first important to establish the relationship between accounting and project controls.

- The relationship is critical to:
  - The overall needs of the project from both the accounting perspective and the actual project development process.
  - The type of accounting and project control systems employed within the organization.
  - The personnel staffing requirements of the organization.
- Accounting controls and project controls are both necessary on all projects, but they serve entirely different needs.

Accounting Controls (*PMfM*, p.448)

- The accounting framework assures financial accountability and:
  - Provides the necessary accounting, audit procedures, and functions.
  - Monitors what has happened regarding expenditures.
- Accounting is therefore part of the mechanism for project control, but it focuses on the results of what happened after-the-fact.
  - Accounting does not measure actual field activity or progress in real time.
  - Accounting analysis gives remaining spend by subtracting actual from budget; then typically and wrongly calculates percent complete by actual spend. This should be avoided.
  - Accounting does not incorporate unknowns or predicted future events (good or bad).
- Accounting does provide the following:
  - Assures that the project's procure to pay process is operational.
  - Manages projects' cash flow through invoice management and payment.
  - Produces monthly financial reports and a historical record of expenditures.
  - Provides protocols to ensure monetary expenditure processes follow accepted practice.

Project Controls (*PMfM*, p.427)

- Project control, when done correctly, is a process that helps steer the project before history is written. Project Controls provides a forward look at upcoming project activity.
- Project control comprises:
  - The effort that ensures the project is correctly defined, estimated, scheduled, and resourced.
  - Significant monitoring of physical progress and field data gathering analytical methods.
  - Monitoring critical trends and forecasting techniques to provide key data for project management to understand and positively influence the project's time and cost outcomes.

Project Control System (PCS) (*PMfM*, p.429)

- PCS is a complete and comprehensive process whereby all aspects of project execution are monitored and then reported against the original approved SOW, budget, and schedule. See Figure 13.1.
- PCS consists of a collaborative collection of processes to support the Project Management Team throughout project execution.
- When combined with suitable software, tools, and the right personnel skill sets, the PCS will provide a definitive integrated source of valuable project information for PMs to use in making informed decisions regarding project scope, cost, and schedule.
- If used appropriately, the PCS can provide an early warning system for the project manager and senior management.

Control Functions Needed On A Project

Scope of engineering, procurement, and construction controls includes:

- Schedule development, updates, and optimization (*PMfM*, p.440)
- Cost Management with budget estimating and value analysis (*PMfM*, p.442/225)
  - Change Management Process that makes use of Potential Deviation Notices (PDNs —trends) (*PMfM*, p.445)
  - Contingency Management (*PMfM*, p.447)
- Performance Measurement of Earned Value and KPIs, with status reporting (*PMfM*, p.413)
- Forecasting of expected results against the FS schedule, including costs and risks
- Risk Management includes register updates, analysis, and mitigation tracking (*PMfM* Chapter 10)
- Contracts management includes progress payments, change orders, guarantees, LDs, warranties, and claims (*PMfM*, p.391).
- Quality control requires the upfront establishment of quality criteria (*PMfM*, p.473)
- Document handling to manage, control, and apply version restrictions (*PMfM*, p.487)
- Accounting controls that are manageable and consistent (*PMfM*, p.448).
- Monetary and currency management controls are in place (*PMfM*, p.450).
- Establish baseline control points for the project. Control documents can include:
  - Scope of Work
  - Detailed cost estimates for work to be performed
  - Project schedule(s)
  - Quality Plan for the SOW to be performed
  - Project goals and objectives clearly defined
- Ensure that procedures such as these are in place:
  - Delegation of Authority
  - Change Control
  - Scope Change Approval
  - Contingency Control
  - Accounting Control
  - Report Frequency

ANALYSIS AND REPORTING

Analysis and reporting provide the Project Manager with the tools to make decisions. See Figure 13.2.

- Utilizing tools, including the PDN log, is necessary for the correct management of contingency and is used to update the Estimate at Completion.
- The weekly report should include a summary of the man-hour burn forecast against actuals.
- Cost updates should show actuals against the approved budget, commitments against budget, and a forecast.
- Analyze progress and report weekly and monthly in concise reports.

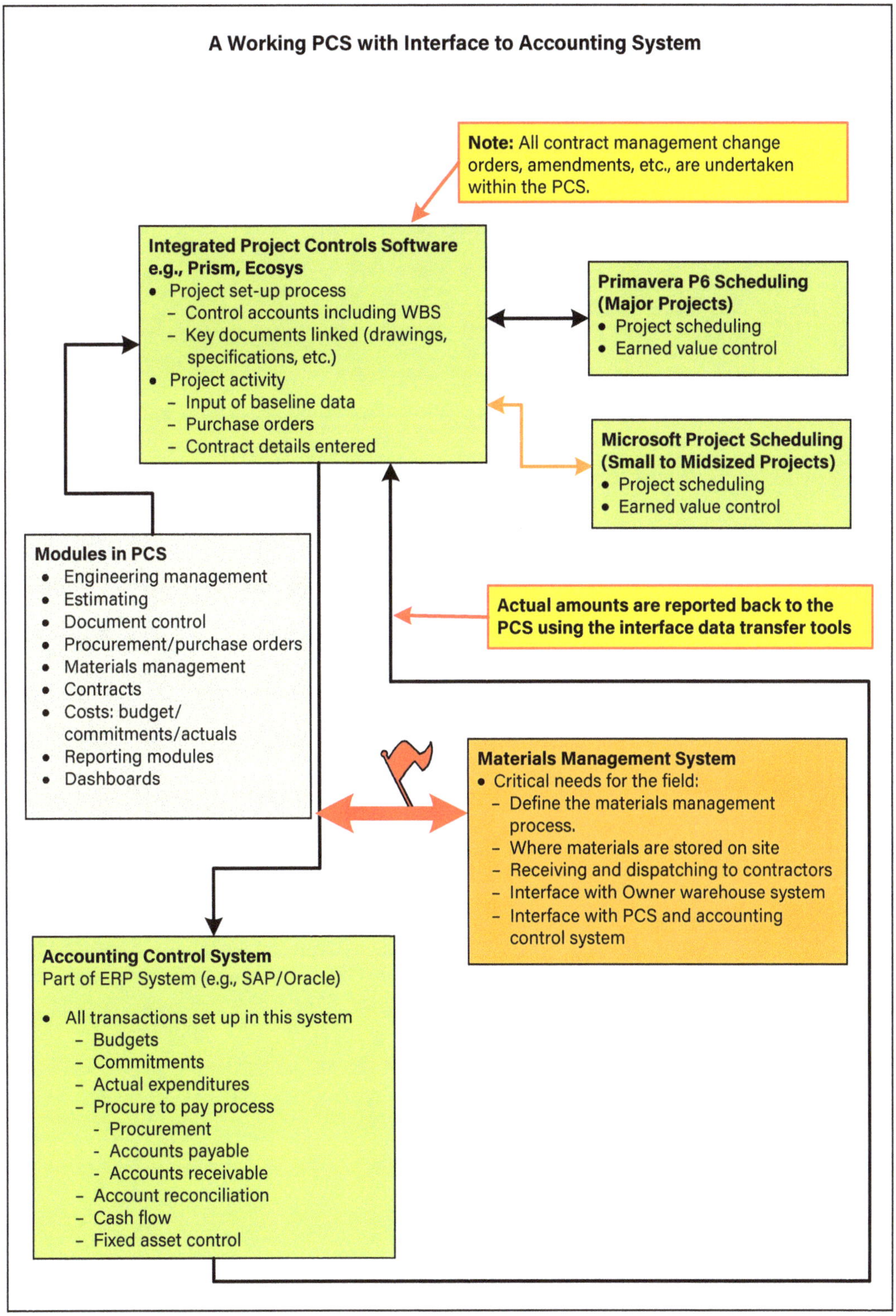

Figure 13.1: Project controls system

	DEVELOP AND EXECUTE PROJECT		MONITOR AND CONTROL PROJECT		
			MONTHLY	WEEKLY	DAILY
PLAN	Establish Project Scope Ch 1,5 Develop Plan for Studies Ch 5 Project Management Ch 1 Process for Achieving Results Responsibility for Process Stakeholder Identification Ch 5 Project Governance Ch 4	Project Execution Plan Ch 18,19 Management Responsibilities Accountabilities Operations Readiness Plan Ch 37 Control Documents Ch 25,27 AFE WBS Schedule/Cost Estimates Project Control System Ch 23,24,26 Project Controls EVM Accounting Controls Quality Personnel	Monthly Planning/Review Rolling 3-Month Project Schedule Cost Labor Safety Initiatives Integrated Project Schedules CPM	Weekly Planning Meeting Set 3-Week Milestones Equipment/Labor Planning Course Correction Milestone Safety 3-Week Construction Schedule	Communication Meetings Work Coordination Issues Productivity Safety
IMPLEMENT	Project Management Select Team Ch 20,21,22 Execute Studies Ch 5,7,8,9,10,11,12,13,14 Scoping Prefeasibility/Feasibility Project Execution AFE Ch 16	Execute Project Setup Project Ch 19 Kickoff Coordination Ch 29 Engineering Ch 30 Procurement Ch 31 Construction Ch 32,33 Commissioning Ch 34,35,36	Review Critical Path Labor Requirements	Review Week's Performance Rocks-in-Road Assessment Weekly Report Activities Concerns Progress KPIs	Set Expectations Coordinate Equipment Labor
MONITOR	Timing and Scope Correction Review Study Results	Periodic Project Report Document	Status of Project Activities Concerns Schedule Cost Critical KPIs Flash Report Monthly Reports	Improvement Plans Drive Implementation Monitor Change Modify Implementation Based on Feedback	Review Past 24 hours
IMPROVE	Management Review Process Ch 6 Executive and Board Approval Ch 17		Review Improvement Plans Value Engineering Activity Rocks-in-Road Issue Priority Rankings		Course Correction Drive Implementation Monitor Obtain Feedback
INVOLVE	President Senior Vice Presidents (VPs) VP/General Manager Project Manager Technical Services Manager	Project Manager Project Team Operating Unit Support Personnel All Employees	Project Sponsor Project Manager Project Team Control Contractors Support Staff	Project Sponsor Project Manager Project Team Area Control Contractors Support Staff	Project Team Contractors All Workers Discipline Leads

Figure 13.2: Project development and control process

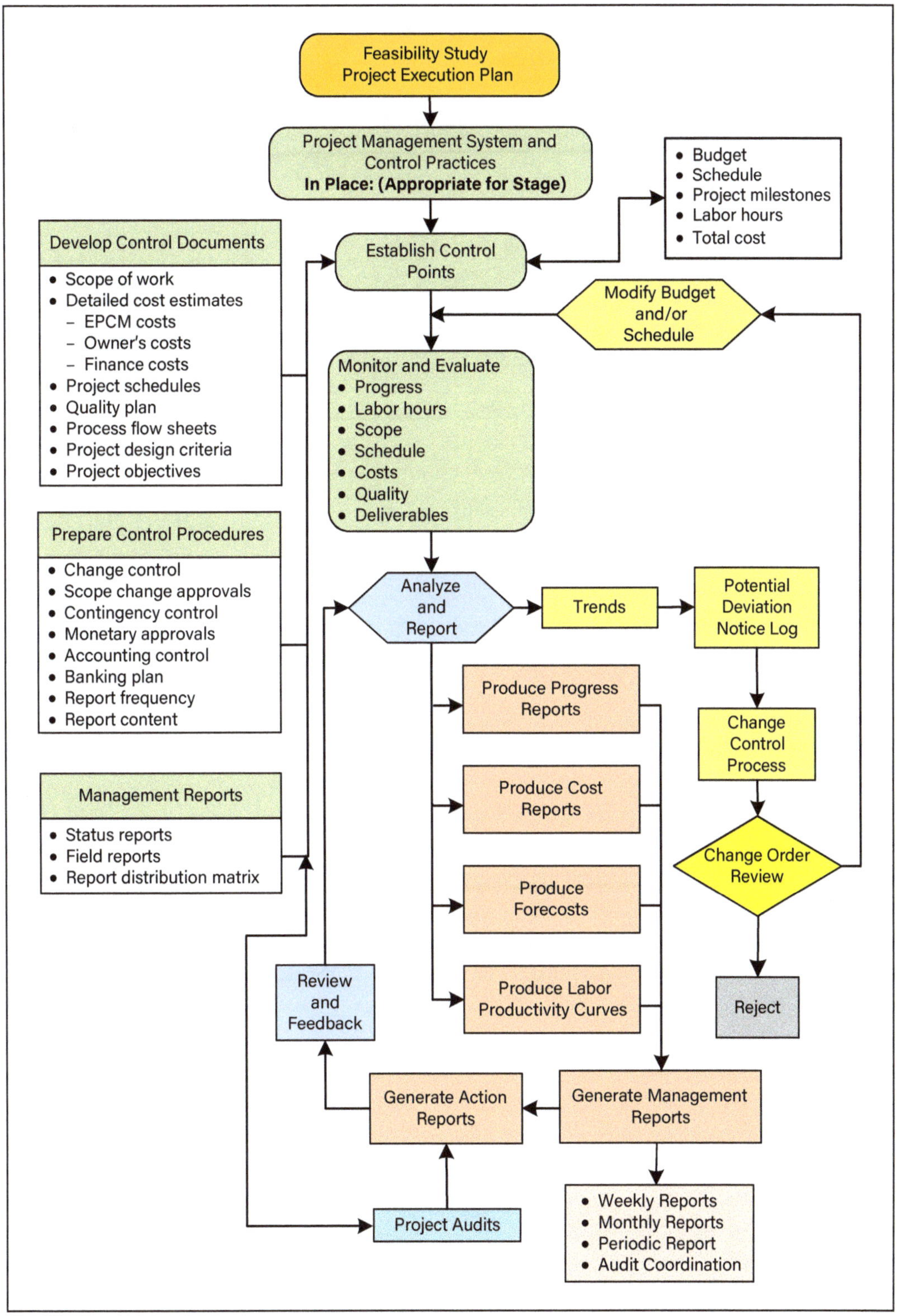

Figure 13.3: Project controls

PROJECT CONTROLS

No.	Item	Status	Date
1	Project management team—in place		
2	Project control management—in place (some may cover multiple duties)		
	A. Controls manager (person responsible for all project costs)		
	B. Estimator		
	C. Scheduler		
	D. Cost engineer (progress measurement, forecasting, etc.)		
	E. Accounting control staff (including audit function)		
3	Control documents—complete		
	A. Approved AFE		
	B. Scope of work—clearly defined and frozen		
	Owner's scope—determined		
	EPCM scope—determined		
	Construction contractor scope—determined		
	C. PEP—finalized		
	D. PPM—assembled		
4	Project control system—developed		
	A. Integrated project control software (IPCS)—installed		
	B. WBS—established		
	Task list and/or activities—completely defined		
	C. Owner ERP system—in place to receive and exchange data		
	D. Project control budget (primary cost control document)—complete		
	E. Cross reference to assets code of accounts—created		
	F. Project baseline schedule—fully detailed and developed		
	G. EVM for measuring progress—utilized		
	H. Change management and control process—in place		
	Major scope change approval process—defined		
5	Project schedule control		
	A. Level 3 project coordination schedule—complete		
	B. Critical path—determined		
	C. Project milestones—all set		
	D. Level 4 execution schedules (for engineering, procurement, and all construction areas)—established		
	E. Operations start-up with construction interface—prepared		
	F. Production schedule (initial 3 years of operation)—produced		

PROJECT CONTROLS (continued)

No.	Item	Status	Date
6	Project procedures and policies—in place		
	A. Project administration		
	B. Accounting procedures—set with Owner		
	C. Contracts management		
	D. Project scope control policy		
	E. Engineering and procurement protocols and control procedures		
	F. Construction management procedures		
	G. Quality management (QA/QC) protocols		
	H. Workforce handbook		
	I. Environmental impact, permits, social acceptance, and sustainability policies		
7	Monetary controls:		
	A. Contingency management, control, and reporting—developed		
	B. Owner's (and financing) cost administration—in place		
	C. Monetary approvals (levels of authority)—established		
	D. Cash flow analysis—undertaken at defined intervals		
8	Management reports—defined		
	A. Report frequency and content—established		
	B. Weekly status, activity, and issues report—defined		
	C. Monthly flash and progress reports—defined		
	D. Periodic project reviews—set		
	E. Dashboard—functioning		
9	Field reports—in place		
	A. Field labor and materials status analysis		
	B. Progress, performance, and productivity measurement		
	C. Field trending		
	D. Personnel projections		
10	Control team functions—defined		
	A. Costs, schedule, quality—monitoring and evaluation parameters set		
	B. Trending of all activities—established		
	C. Forecast updates of project costs and schedule—determined		
	D. Owner and management reporting basis—set		
	E. Project compliance analysis—determined with Owner		
11	Audit processes—established		
	A. Project audits (for adherence to best practices)—set		
	B. Financial audits—frequency established		
	C. Project management audits—undertaken, if appropriate		
12	Recovery plan requirements—followed as necessary (*PMfM* Chapter 28)		

SECTION 14

Operations Readiness

OBJECTIVE

A project is not complete until it is up and running, attains commercial production, and is on its way to fully achieving all stated operational goals. The Operations Readiness Plan (ORP) is a document that spells out how the project will be phased into the operating business.

Each section of these guidelines is designed to direct the reader to the corresponding chapter in *PMfM*, which must be thoroughly understood for successful implementation. *PMfM* must be used alongside these guidelines.

REFERENCES

Item	*PMfM*	Page
Introduction	Chapter 1	3
Ethics	Chapter 3	23
Governance	Chapter 4	33
Scoping Evaluation	Chapter 5	49
Operations Readiness	Chapter 37	679

CG Figure	Figure Caption	*PMfM* Reference
Figure 14.1	Operations readiness	Flowchart 37.1, p.696
None	Operations readiness by overall property	Figure 37.2, p.682
None	Operations readiness department status	Figure 37.5, p.688

Checklist Title	*PMfM* Reference
Operations readiness	Checklist 37.1, p.694

SCOPE

A crucial element in achieving the project's desired "end result" is the operating team that will take it forward. Key to successful operations deployment is a well-defined ORP that is implemented in a timely manner and properly managed.

- When do you need this operating team?
- What does the operating team need to know?
- Who do you need in the operating team, and in what order are they needed?

Key Elements of the Operations Readiness Plan

- Methods for phasing the project into an operating entity.
- An acquisition plan for the people, training, systems, processes, and services required to start, run, and maintain the operating facility. This requires detailed position skill requirements and the recruitment staffing schedule.
- A plan for timely development of operational capability.
  - Identification of interfaces between contributing parties, including OEMs
  - Arrangements for progressive roster ramp-up and plant handover (Figure 14.1)
- Requirements for business performance:
  - Safety plans and monitoring controls in place.
  - Regulatory compliance policies, procedures, and tracking systems ready.
  - Quality standards documented and quality control measures in place.
  - Community and Social Responsibility action plans and responses established.
- Identification of risks and opportunities during execution.
  - Consider giving preference to local hires. Have the discipline to release contractor personnel at project end.
  - Failure to transition to operations teams is costly and may be driven by contractors if not actively managed by the project and operations teams.
- Operations Readiness feedback tools include tables, charts, and other metrics based reporting tools (*PMfM* Figures 37.2 and 37.5).

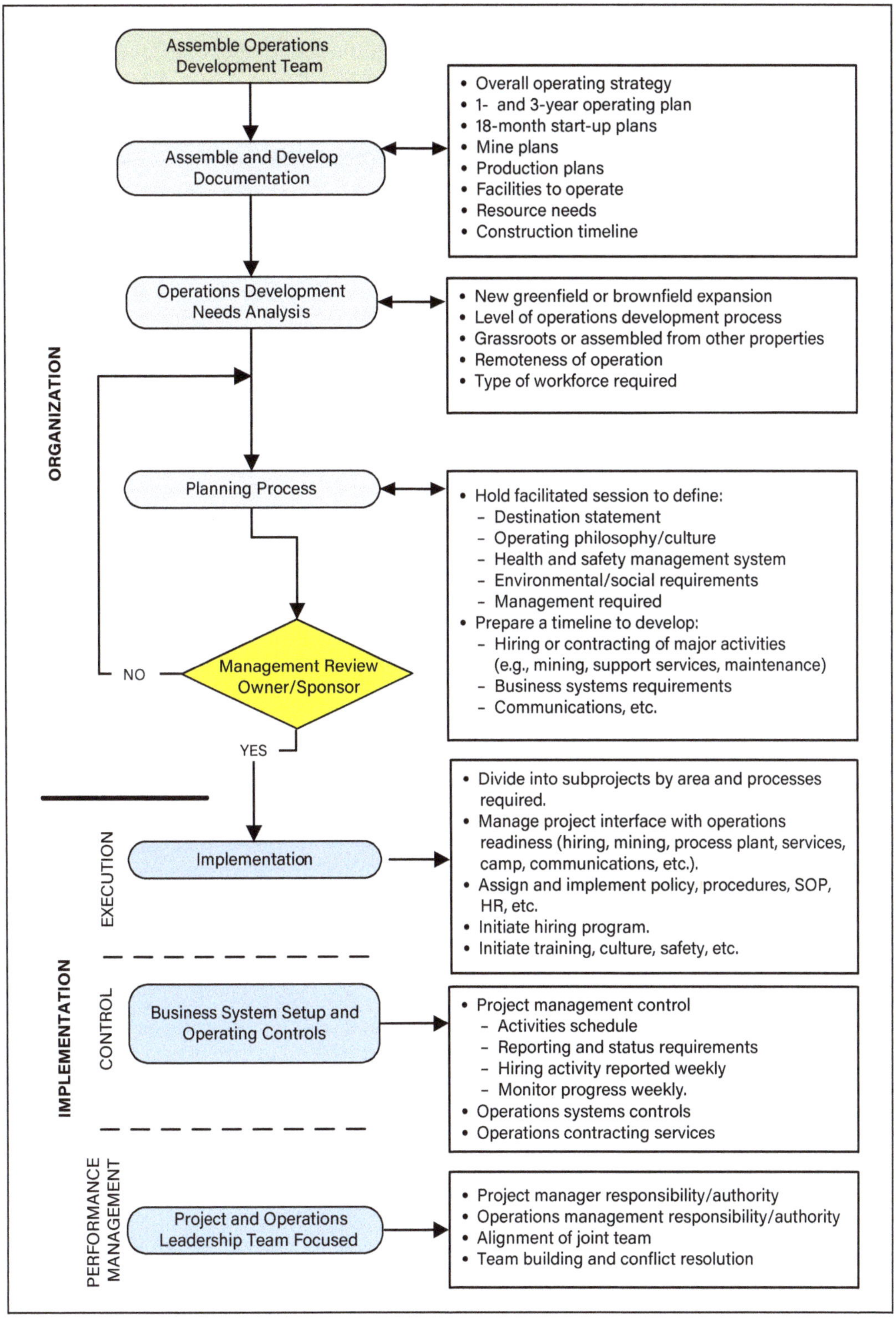

Figure 14.1: Operations readiness

OPERATIONS READINESS

No.	Item	Status	Date
1	Strategy for development of operational capability—established		
	A. Start-up year objectives, including realistic mine and plant ramp-up targets for tons, grade, and recovery—finalized		
	B. Minimum requirements—agreement reached with Owner		
	C. Implementation timing—set with Owner's operations team		
	Operational capability development to be complete 3 months before precommissioning takes place		
2	Core operations management team—recruited and in place in a timely way		
3	ORP—developed		
	A. ORP elements aligned to match project start-up goals		
	Operations readiness programs for all departments—set		
	B. PEP—in hand (for schedule alignment)		
	Project commissioning and start-up plan—available for reference		
	C. Operations plan (annually for initial 3 years)—complete		
	Production systems requirements—specified		
	Maintenance plan and tooling requirements—set		
	D. Organization recruitment program—approved by Owner		
	Organizational structure—finalized		
	Hiring team—in place with a defined timeline		
	Employee handbook (for operations)—published		
	E. Organizational training program—prepared		
	Operations training materials—developed		
	Operations training—conducted ahead of start-up		
	Operator and maintenance handbooks—available to site personnel for every piece of equipment		
	Vendor support for commissioning/start-up—arranged		
	F. IT systems to support operations—in place		
	G. Business system/ERP system requirements—identified		
4	Department and function plans, SOPs, and policies—developed		

OPERATIONS READINESS (continued)

No.	Item	Status	Date
5	Individual departmental readiness plans—prepared		
	A. Geology		
	B. Mine engineering, operations, and maintenance		
	C. Metallurgy and processing		
	Training for the process control systems—arranged		
	D. Human resources and training		
	E. Health and safety, and loss prevention		
	F. Security systems—taken over from the project		
	G. Procurement		
	Warehousing needs—established		
	First fills and 2 years of spares—in place and cataloged		
	H. Information technology		
	I. Environmental, permits, and social responsibility		
	Resources to implement the obligations—identified		
	Regulatory agency approval requirements—known		
	Responsibilities—assigned		
	J. Accounting		
	K. Legal		
	L. Operations management		
6	Support services requirements—services agreements identified		
7	Project and operations alignment session—conducted		
	A. Culture establishment—facilitated by project manager (PM) and Owner		
8	ORP links and dependencies between functions—identified		
	B. Communications between functions—facilitated by PM		
9	Operations risk and opportunities review—conducted		
10	ORP status reporting—initiated at project execution kickoff		
	A. Program development monitoring—responsibility of PM		
	B. Readiness check reviews—conducted at set intervals		
	C. ORP changes—authorized by PM (with Owner approval)		

SECTION 15

Project Set Up

OBJECTIVE

This section defines all activities required following completion of the FS and before project execution commences.

This section provides the methodology for the project manager to take the PEP and set up the project in accordance with project management best practices. The project execution clock starts when the official green light (the AFE) is issued. This chapter captures the planning steps that will give execution a strong start.

The various disciplines required of the project team are outlined here. *PMfM* Chapter 19 presents the project philosophy with respect to engineering, procurement, and construction. The documents and resources needed to initiate project execution procedures are described, along with the informational reports and meetings that key stakeholders will be expecting.

Each section of these guidelines is designed to direct the reader to the corresponding chapter in *PMfM*, which must be thoroughly understood for successful implementation. *PMfM* must be used alongside these guidelines.

REFERENCES

Item	***PMfM***	**Page**
Project Setup	Chapter 19	299
Environment, Permits, Social, and Sustainability	Chapter 9	113
Executive and Board Decision	Chapter 17	271
Project Execution Plan	Chapter 18	277
Team Organization	Chapter 20	323
Project Deliverer	Chapter 21	353
Project Controls	Chapter 24	427
Reports and Progress Reviews	Chapter 25	457
Quality Assurance and Quality Control	Chapter 26	471
Document Management and Document Control	Chapter 27	487
Engineering	Chapter 30	523
Procurement	Chapter 31	571
Construction	Chapter 32	593
Precommissioning	Chapter 34	637

CG Figure	**Figure Caption**	***PMfM* Reference**
Figure 15.1	Communicating project complexity	Figure 19.5, p.315
Figure 15.2	Project setup	Flowchart 19.1, p.322
Figure 15.3	Kickoff coordination	Flowchart 29.1, p.522

Checklist Title	***PMfM* Reference**
Project set up	Checklist 19.1, p.321
Kickoff coordination	Checklist 29.1, p.521

SCOPE

There are several key areas that ensure the project is set up to execute the PEP.

- Project team organization and selection of the PM should be done according to project requirements (*PMfM* Chapter 17). Project size (CAPEX) will guide team size, but all aspects need to be covered.
- A properly completed PEP will provide the guidance for finalizing the project strategy.
- Plan how to implement the project, whether the project is in the FS, engineering and procurement, or construction phase.
- Establish the necessary Project Controls (*CG* Section 13 and *PMfM* Flowchart 24.1, p.456).
- Begin the reporting process and agree on the types and frequency of reports (*PMfM* Chapter 25).
- The QA/QC program is to be agreed upon before work commences; this includes the reporting format and frequency (*PMfM* Chapter 26).
- Document management establishes the necessary management and control processes (*PMfM* Chapter 27).
- Selection of contractors will be significantly different in the execution stage from the Viability Stage. Pay close attention to the process, including the RFP process and key deliverables in the bids. Selection is very different for project work than it is for an operations contracting process.
- Contract completion for a full contract with an EPCM firm and specialist engineers should be part of the organization plan (*PMfM* Chapter 21).
- Project kickoff is an important event at the start of any project, regardless of stage, to ensure that all key groups and people are aligned and agree on how they will work together.
- Project implementation is now active and the team is working the plan. Management control is key to keeping coordination and issue resolution a priority.
- Project leadership needs to comprehend the complexity of executing a major mining project. Figure 15.1 highlights the interactions that take place, and the context is explained in *PMfM* Chapter 19.

Kickoff

The Project Manager's role is to ensure everyone is aligned on objectives. This is accomplished in a formal kickoff meeting. The kickoff meeting ensures a shared commitment by key players to deliver a project that meets all stated goals (*PMfM*, p.513).

- The team is brought together when engineering and construction efforts commence.
- The alignment (partnering) session clearly identifies the roles and relationships among the players.
  - Conflict resolution protocols are established (but the PM still has final authority).
- Environmental and social representatives explain commitments that the team must uphold.
- Performance measures are set and confirmed.
  - Reporting requirements, meetings, and audit protocols are established during the kickoff meeting.
- Confirm that key project procedures *are* in place and functioning.

Project Procedures Status Levels

Below is a list of Project Procedures Status Levels with associated *PMfM* chapters.

- PEP is prepared with SOW fixed and a WBS adopted (*PMfM* Chapter 18, p.279).
- Organization chart is established (*PMfM* Chapter 20, p.329).
- Project controls are functioning (*PMfM* Chapter 24, p.428).
- Engineering protocols and procedures are set (*PMfM* Chapter 30, p.527).
- Procurement protocols and procedures are set (*PMfM* Chapter 31, p.571).
- Construction Management Plan is outlined (*PMfM* Chapter 32, p.602).
- Commissioning Plan is established and a completion date is set (*PMfM* Chapter 34, p.638).
- Environmental and permitting requirements are documented (*PMfM* Chapter 9, p.119).
- Social acceptance and sustainability needs and roles are understood (*PMfM* Chapter 9, p.122).
- Project administration personnel and procedures are in place (*PMfM* Chapter 19, p.309).
- Communications Plan is settled (*PMfM* Chapter 19, p.312).
- Reporting requirements are decided (*PMfM* Chapter 19, p.316).

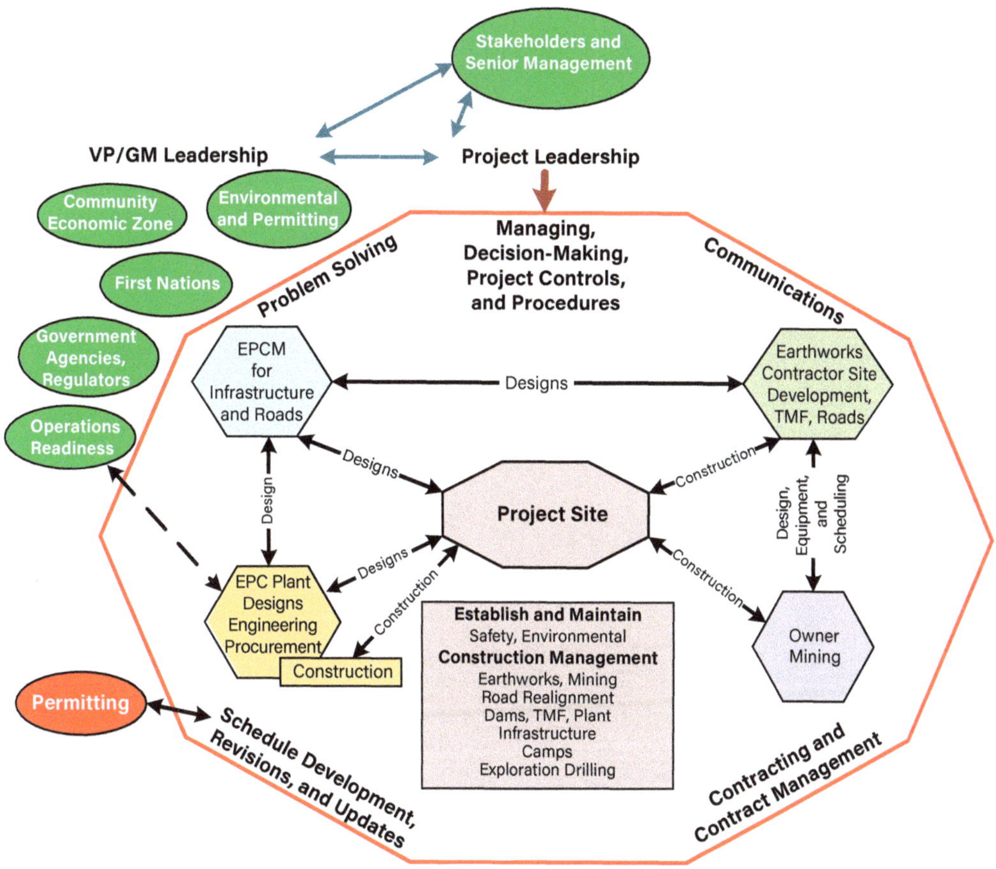

Figure 15.1: Communicating project complexity

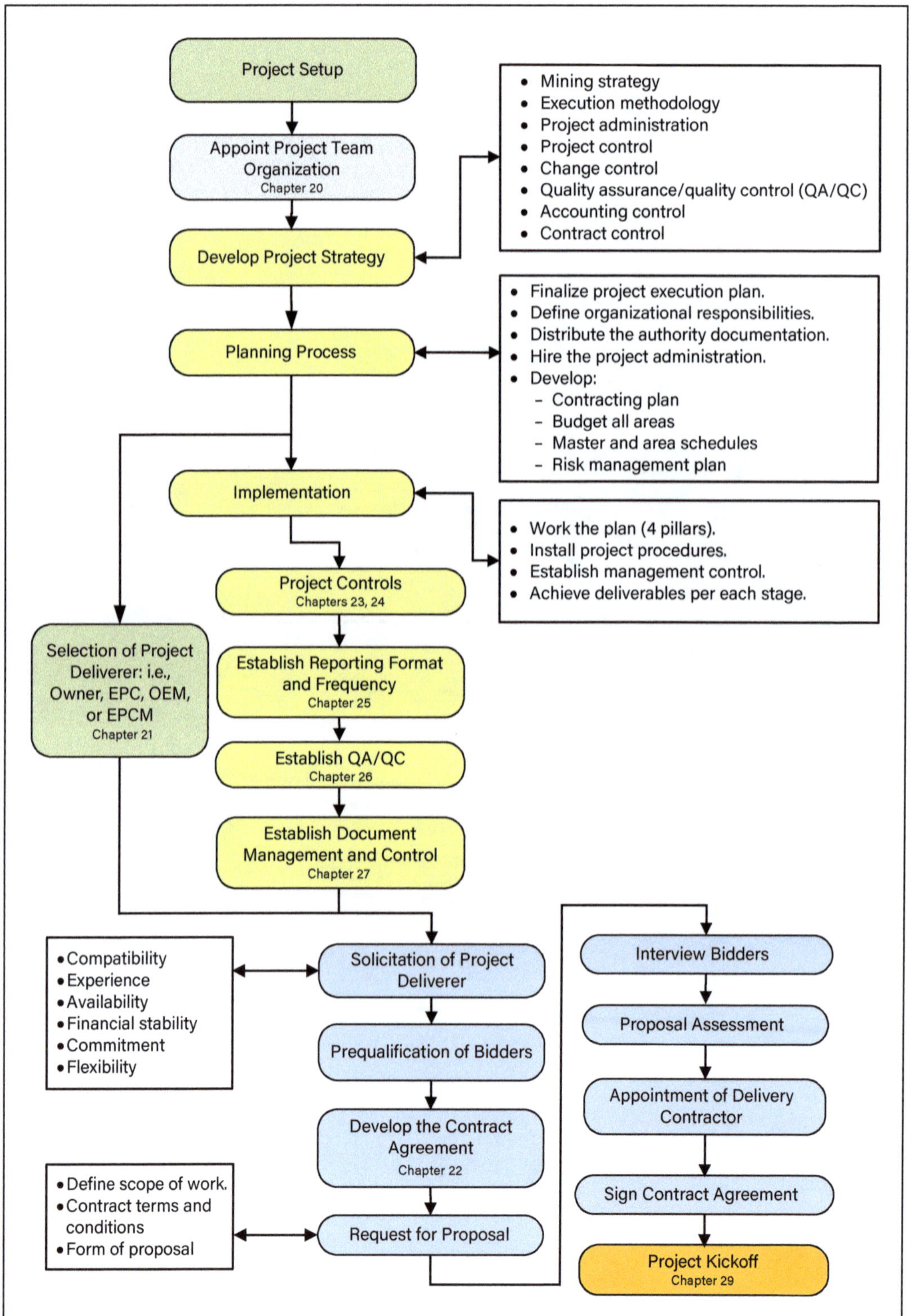

Figure 15.2: Project setup

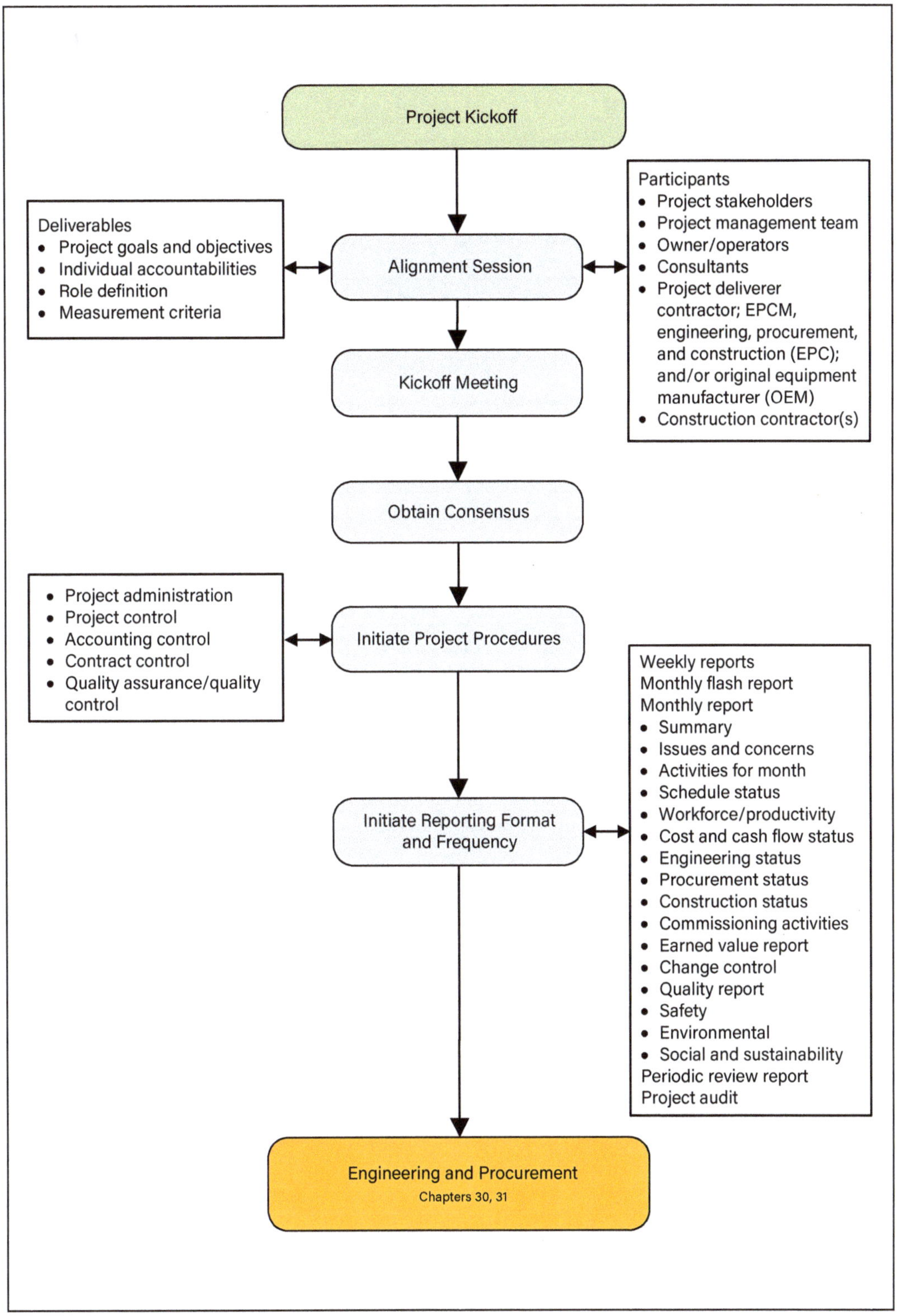

Figure 15.3: Kickoff coordination

PROJECT SET UP

No.	Item	Status	Date
1	Owner's management approval for project execution—obtained		
2	Project goals—identified		
	A. Destination statement—published		
	B. Stakeholder expectations—understood and aligned		
	C. Deliverables—documented		
3	Project governance—established		
	A. Corporate leadership role—clarified		
4	Project manager (PM)—appointed		
	A. PM empowered with the authority to execute the project—OK		
	B. Clear line of authority from PM to corporate leadership—set		
5	Project team—selected		
	A. Appropriate team for project type and size—selected		
	B. Team responsibilities—assigned		
	C. Administration support staff—assembled		
6	Project strategies and philosophy—delineated		
	A. Owner's risk threshold—understood		
	B. Contracting approach—established		
7	Risk assessment—undertaken		
	A. Comprehensive risk mitigation plan—in place		
8	PEP modified to reflect approval terms—complete		
	A. Final PEP—in place		
	B. Scope of work—refined and frozen		
	C. WBS—established		
	D. Project delivery provider selection process—determined		
9	Project procedures—set		
	A. PPM—assembled		
	B. Project administration procedures—in place		
	C. Engineering protocols and procedures—established		
	D. Procurement (sourcing and logistics) procedures—set up		
	E. Construction management procedures—prepared		
	F. Document management and document control—established		
	G. Project control process and project system—set up		
	H. Change control system and procedures—in place		
	I. Scope control policy—set, as a part of project governance		
	J. Quality management plan—functioning		
	K. Communications plan—implemented (internal and external)		
	L. Insurance requirements—satisfied		
	M. Environmental and permitting checklists—prepared		
	N. Social program requirements—understood		
10	Informational technology resources—in place		
11	Reporting and information feedback system—established		
	A. Weekly flash timing—fixed		
	B. Monthly progress report contents—set		
	C. Field reports—set up		
	D. Dashboard—functioning		
	E. Third-party external reviews—available		

KICKOFF COORDINATION

No.	Item	Status	Date
1	Project team—assembled		
2	Preparation for kickoff and alignment session—complete		
	A. External facilitator—engaged		
	B. Agenda for kickoff meeting—published		
3	Relevant attendees—identified by name		
	A. Project management team		
	B. Project stakeholders		
	C. Key Owner personnel and operators		
	D. Key EPCM members (including sponsor)		
	E. Major construction contractors		
	F. Commissioning team leadership		
	G. Crucial specialists, consultants, and vendors		
4	Partnering or alignment session—held per guidelines		
	A. Project vision statement—produced		
	B. Project vision statement—signed by all		
	C. Project goals—consensus achieved		
	D. Individual roles and responsibilities—defined		
	E. Alignment deliverables—identified		
	F. Conflict resolution—addressed		
5	Partnering session feedback—incorporated into PEP		
6	Project procedures—implemented		
7	Environmental and permitting requirements—understood		
8	Social acceptance and sustainability needs—understood		
9	Reporting requirements and distribution lists—documented		
	A. Weekly and monthly report formats—established		
	B. Issue date and responsibility for each report—set		
10	Meetings purposes and frequency—established		
11	Audit program—initiated		
12	Follow-up alignment sessions—dates established		

SECTION 16

Execution: Engineering, Procurement and Construction

OBJECTIVE

With the project feasibility study completed, the AFE and board approvals are received, and all the activities outlined in *CG* Section 15 are complete. The project is then ready to move forward into the execution stage. This Section outlines the engineering, procurement and construction activities.

Each section of these guidelines is designed to direct the reader to the corresponding chapter in *Project Management for Mining (PMfM)*, which must be thoroughly understood for successful implementation. *PMfM* must be used alongside these guidelines.

REFERENCES

Item	*PMfM*	Page
Engineering	Chapter 30	523
Procurement	Chapter 31	571
Construction	Chapter 32	593

CG Figure	Figure Caption	*PMfM* Reference
Figure 16.1	Engineering life cycle	Figure 30.5, p.566
Figure 16.2	Engineering	Flowchart 30.1, p.569
Figure 16.3	Procurement	Flowchart 31.1, p.591
Figure 16.4	Construction	Flowchart 32.1, p.623

Checklist Title	*PMfM* Reference
Engineering	Checklist 30.1, p.567
Procurement	Checklist 31.1, p.589
Construction	Checklist 32.1, p.621

SCOPE

Detail Engineering

This section focuses on the detailed engineering required to produce Issued For Construction (IFC) drawings and the related procurement information. Engineering from the FS should have advanced to plus 40% with basic engineering at 85%. The engineering phases to reach this point are discussed in *PMfM* Chapter 30 and the reader is encouraged to read that chapter carefully to understand how it advances through the project stages. The engineering work phases (*PMfM* p.527) include:

- Predesign phase to establish the Engineering Work Program
- Site Analysis Study which includes data gathering and evaluation
- Conceptual Design initiation which establishes Owner requirements
- Basic (Design) Engineering which sets the functional philosophy. Basic engineering sets the initial PFDs, specs, P&IDs, plot plans, equipment lists and required standards.
- Detail (Final) Engineering is where the IFC drawings are delivered
- Drawing Reviews occur throughout basic and detail engineering
- Specifications development parallels basic and detail engineering efforts
- Engineering scope control which applies to all engineering work phases
- Deliverables of detail (first design) engineering (Figure 16.1)
- Cost-estimating for budget and construction bidding support (*PMfM*, p.555)
- Construction bidding support and contract agreement award support
- Field Engineering is a site support role for constructor and Owners (*PMfM*, p.557)
- Project completion and closeout (*PMfM* Chapter 33 and 36)

Procurement

Procurement includes the following steps (*PMfM*, p.579):

- Creation of a clear sourcing procurement road map
- Segment goods into work packages. The packages must recognize market realities.
- Ascertain the qualified suppliers and identify long lead time equipment
- Pursue a cost minimization and efficient contracting strategy
- Beneficially align EPCM and Owner procurement organizations
- Purchasing generates formal bids though Request for Proposal (RFP). Bids and proposals are reviewed and Purchase Orders (PO) or Contracts are awarded (*PMfM*, p.580).
- Expediting requires creating a plan to deliver goods timely on site (*PMfM*, p.582)
- Logistics team or group needs to know how to move and track materials from source to site. Use a freight forwarder and a customs agent (*PMfM* Figure 31.1, p.586)

Construction (*PMfM*, p.602–608)

Construction activities include the following:

- Construction and battery limits set in logical work packages
- Long lead time items are all identified and ordered in time
- Prime construction contractor(s) and key sub-contractors selected and onboarded
- Master schedule (AACE Level 3) is in place with all constructors aligned
- Engineering complete to appropriate level with >70% of drawings at IFC
- Construction Plan is in place with sequences, methods, and interfaces captured
  - DO NOT MOBILIZE TOO EARLY. The project will need multiple ongoing work fronts to handle large construction teams.
  - Quality, safety, permitting, and Environmental and Social programs are setup and functional with appropriate site staff.
- Construction workforce size and buildup rate determined
- PM team in field, completely focused on construction of facilities
- Seasoned Construction Manager and professional Contracts Administrator on Project Management team
- Progress report format, Earned Value Management accepted by all constructors

Construction Contractor Selection (*PMfM*, p.594)

- Bid Invitation follows a Request for Quotation (RFQ) that removes the unqualified bidders. Include SOW, Level 2 schedule, bond and reporting requirements (*PMfM*, p.195).
- Bid Evaluation issues for the PM to review (*PMfM*, p.596)
  - Is the Compliance to the SOW outlined in RFP?
  - Are technical requirements conformance including QA/QC obligations addressed?
  - Review of contractor's past performance and current capability
  - Review exceptions taken by contractor and the alternatives offered
  - Bid price evaluation complete to show like-for-like comparison
- Bid Assessment matrix grid scorecard completed with typical weighting ranges included (*PMfM* Figure 21.3, p.377):
  - Personnel and technical competency [25–37%]
  - Plan of execution including contract strategy [35–55%]
  - Contractor performance assessment including safety [10–22%]
  - Other category (e.g., cost competitiveness, permitting prowess) [4–12%]

The Project Management Role

Project Management within the construction phase includes:

- Work Strategies
- Contractor Selection
- Duties and Tasks of Project Management personnel
  - Project Manager (PM)
  - Construction Manager (CM)
  - Field Engineer
  - Contracts Manager
- The Construction Plan
- Key Elements
- Enhancement Techniques
- Construction Completion (*PMfM* Chapter 32)
  - What constitutes contract completion?
  - Completion definitions and their differing ramifications

PM Duties during Construction (*PMfM*, p.594, 600–602, 608–618)

- Ensure Owner selects competent constructors. Obtain executed contracts with bonds and insurance certificates.
- Select and bring in adequate qualified staff for site Project Management team. Enforce a safe work environment by establishing quality standards for materials and work.
- Integrate engineering and procurement with construction from the project outset.
  - Conduct the preconstruction kickoff. Set a workable IFC drawings sequence timetable.
  - Implement site materials management and arrange external transport and goods deliveries.
- Arrange for permits to be in place in timely manner
- Manage project risk by enacting strategies to reduce and mitigate risks. Install site security system and ensure labor relations issues do not impede work progress.
- Assert sufficient control to ensure schedule is met and facilities are completed on budget. Maximize work fronts by minimizing site constraints while maintaining site utilities, access and facilities.
- Provide work-site services including site access, offices, utilities, and temporary facilities
  - Manage internal site movements and coordinate crane lifts. Work to level the site support equipment usage to limit total number of support equipment required.
  - Implement site communications plan and install document handling system
- Monitor productivity by measuring and reporting work progress. Check that quality is being met. Provide contract interpretations, approve all changes and authorize progress payments.
- Prepare for precommissioning and operations start-up

Winning Techniques for Construction

- Acquire as many permits in advance as possible
- Set construction calendar to minimize the impact of inclement weather
- Improve access roads and relocate utilities out of harm's way
- If budget allows in advance of construction start:
  - Place as many temporary facilities as authorities will allow
  - Set up preliminary infrastructure such as power, communications, etc
  - Locate and acquire sources of construction materials and potable water
  - Undertake site grading, fencing and drainage modifications
  - Install water wells, batch plant, powder magazines, and survey points
  - Construct warehouses, laydown yards and camp facilities
- Maximize offsite fabrication using pre-assembly and modular items
- Do not mobilize civil contractors until sufficient work fronts are ready. Generally the project should have more than 75% of civil drawings at IFC, and contract is executed (4D modeling in *CG* Section 9 can assist with this timing).
- Remember to perform all activities with the end in mind!

ENGINEERING—LIFE CYCLE FOR A PROJECT

STUDY	SCOPING	PREFEASIBILITY with TRADE-OFFS		FEASIBILITY	
ENGINEERING	PRELIMINARY—CONCEPTUAL DESIGN				
			BASIC (DESIGN) ENGINEERING		
	5%	10%	20%	25%	35%
	3-D MODEL PROGRESS			3-D MODEL PROGRESS	
	• PRELIMINARY MODEL	• MAJOR EQUIPMENT • BUILDING(S) ENVELOPE • ACCESS ROADS/BENCHES • SITE (PIT, TSF, DUMPS) • TRADE-OFF STUDIES	• OPERATING SPACES • MAINTENANCE AREAS • LAYDOWN YARDS • MAJOR UTILITIES/TIE-INS • PRELIM CIVIL MODEL	• EQUIPMENT LOCATIONS • MAJOR ARCHITECTURAL • BASIC STRUCTURAL STEEL • BASIC CONCRETE LAYOUT • ELEC CONTROL ROOM	• EQUIPMENT MODEL • BUILDINGS—with Stairs, Platforms, Containments • PROCESS PIPE MODEL • LARGE DIAMETER PIPING
	DELIVERABLES			DELIVERABLES	
	PLANT LAYOUT • BLOCK FLOW DIAGRAM	PLANT LAYOUT • PRELIM FLOWSHEET • PRELIM SITE LAYOUT	PLANT LAYOUT • PROCESS FLOWSHEET • PRELIM SITE & PLOT	PLANT LAYOUT • PRELIM OVERALL GAs	PLANT LAYOUT • CONVEYOR PROFILES • SITE & PLOT PLANS
SUPPORTING DOCUMENTS	PROCESS • BLOCK FLOW DIAGRAM • INITIAL PROCESS DESIGN • PRELIM MASS FLOW	PROCESS • METSIM MODEL • MASS BALANCE • OPERATING PROCEDURE	PROCESS • DESIGN CRITERIA • METALLURGICAL DATA • MASS & WATER BALANCE	PROCESS • MET TEST PROGRAM • DESIGN CRITERIA	PROCESS
	CIVIL • SITE TOPOGRAPHY	CIVIL • SITE TOPO & FEATURES • GEOTECH REPORT • DESIGN CRITERIA	CIVIL	CIVIL • ROADWAYS & BENCHES • DITCHES & CULVERTS • RETENTION BASINS	CIVIL
	MECHANICAL • PRELIM EQUIP SIZING • SITE CONDITIONS	MECHANICAL • MECHANICAL EQUIP LIST • MAJOR EQUIP SIZE—TYPE • ENGINEERING CODES	MECHANICAL • MAJOR PIPELINE SIZING • TANK SIZING • MAJOR EQUIP SPECS	MECHANICAL • REC'D BIDDERS LIST • MECH EQUIP SPECS • UPDATED EQUIP LIST	MECHANICAL • P&ID INPUT • PIPELINE SIZING • COMPONENTS SIZING
	STRUCTURAL	STRUCTURAL • DESIGN CRITERIA • BUILDING CODES	STRUCTURAL	STRUCTURAL	STRUCTURAL
	ELECTRICAL	ELECTRICAL • DESIGN CRITERIA • ELECTRICAL CODES • ETAP MODEL	ELECTRICAL • ONE-LINE DIAGRAMS • UTILITY LOCATIONS • PRELIM ELEC EQUIP	ELECTRICAL • ELECTRICAL EQUIP LIST • DEFINED ELEC ROOMS • POLE LINE UTILITY ROUTE	ELECTRICAL
	PROCESS PIPING	PROCESS PIPING • DESIGN CRITERIA	PROCESS PIPING	PROCESS PIPING • PRELIMINARY P&ID	PROCESS PIPING • P&ID WITH LINE SIZING • MATERIAL & VALVE SPECS
	INSTRUMENTATION	INSTRUMENTATION	INSTRUMENTATION	INSTRUMENTATION	INSTRUMENTATION
	PROJECT MANAGEMENT • SITE VISIT REPORT • SCOPE OF WORK	PROJECT MANAGEMENT	PROJECT MANAGEMENT	PROJECT MANAGEMENT	PROJECT MANAGEMENT
	CLIENT • PROJECT DESCRIPTION	CLIENT • SITE & RESOURCE DATA • CONSULTANT INFO	CLIENT • SITE SURVEY	CLIENT • CONSULTANTS (Geology, Mine, TSF, Enviro, Hydro)	CLIENT • CONSULTANTS (Geology, Mine, TSF, Enviro, Hydro)

Figure 16.1: Engineering life cycle

ENGINEERING—LIFE CYCLE FOR A PROJECT

FEASIBILITY	DEFINITIVE ESTIMATE

FINALIZE BASIC ENGINEERING

DETAIL ENGINEERING

45%	55%	70%	85%	95%
	3-D MODEL PROGRESS		**3-D MODEL PROGRESS**	
• CONCRETE FOUNDATIONS • VENDOR EQUIP UPDATE • CIVIL GRADING MODEL • PREMIUM EL' EQUIP MODEL • PRELIM PIPING MODEL	• STRUCTURAL MODEL • CONCRETE MODELS • PROCESS PIPING MODEL • ELEC RACEWAY MODEL • ARCHITECTURAL MODEL	• MECHANICAL DETAILING • FINAL PIPE ROUTING • STEEL PLATFORMS • FINAL CONCRETE MODEL • ELEC. EQUIPMENT MODEL	• FINAL VENDOR MECH/ELEC • BUILDING DESIGN UPDATE • FINAL PIPING MODEL • FINAL STEEL MODELS • FINAL CONCRETE MODEL	• EQUIP LAYOUT COMPLETE • FINAL MISC. STEEL MODELS • MISC CONCRETE MODEL • FINAL MISC. ELEC. MODELS • FINAL HVAC COMPONENTS
	DELIVERABLES		**DELIVERABLES**	
PLANT LAYOUT • FINAL FLOWSHEETS • GAs by PROCESS AREA	**PLANT LAYOUT** • MECH DETAIL DRAWINGS ISSUED FOR FABRICATION	**PLANT LAYOUT** • GAs ISSUED FOR DESIGN	**PLANT LAYOUT**	**PLANT LAYOUT**
PROCESS	**PROCESS**	**PROCESS**	**PROCESS**	**PROCESS**
CIVIL • ROUGH GRADING MODEL • TOPO MODEL UPDATE	**CIVIL** • FINAL GRADING MODEL • SEDIMENT CONTROL PLANS	**CIVIL** • SEDIMENT CONTROL • EROSION CONTROL	**CIVIL** • COMPLETED CIVIL MODEL	**CIVIL**
MECHANICAL • VENDOR CUTSHEETS	**MECHANICAL** • MAJOR EQUIP VENDOR INFO	**MECHANICAL** • FINAL MECH EQUIP—VENDOR INFORMATION • FIRE PROTECTIONS	**MECHANICAL** • FINAL P&ID LINE/VALVES • HVAC/BUILDING SERVICES • PLUMBING	**MECHANICAL**
STRUCTURAL • MATERIAL TAKE-OFFS	**STRUCTURAL** • STEEL FRAMING DATA • FOUNDATION ENGRG	**STRUCTURAL** • STEEL STRUCTURES • CONCRETE FOUNDATIONS	**STRUCTURAL** • ACCESS PLATFORMS, STAIRS • CURBS, SLABS	**STRUCTURAL** • STEEL for MINOR EQUIP • MINOR EQUIP CONCRETE
ELECTRICAL	**ELECTRICAL**	**ELECTRICAL** • FINAL ELECTRICAL EQUIP—VENDOR MODELS • LIGHTING	**ELECTRICAL**	**ELECTRICAL**
PROCESS PIPING • TANK DATA SHEETS • LINE & VALVE LISTS	**PROCESS PIPING** • APPROVED P&IDs • SPECIALTY ITEMS	**PROCESS PIPING** • PIPING MODEL • PIPE RACK STEEL	**PROCESS PIPING**	**PROCESS PIPING**
INSTRUMENTATION	**INSTRUMENTATION**	**INSTRUMENTATION**	**INSTRUMENTATION**	**INSTRUMENTATION**
PROJECT MANAGEMENT	**PROJECT MANAGEMENT**	**PROJECT MANAGEMENT**	**PROJECT MANAGEMENT**	**PROJECT MANAGEMENT**
CLIENT • CONSULTANTS (Geology, Mine, TSF, Enviro, Hydro)	**CLIENT** • CONSULTANTS (Geology, Mine, TSF, Enviro, Hydro)	**CLIENT** • CONSULTANTS (TSF, Enviro, Hydro)	**CLIENT** • CONSULTANTS (TSF, Enviro, Hydro)	**CLIENT** • CONSULTANT (TSF, Environmental, Hydrology)

Figure 16.1: Engineering life cycle (continued)

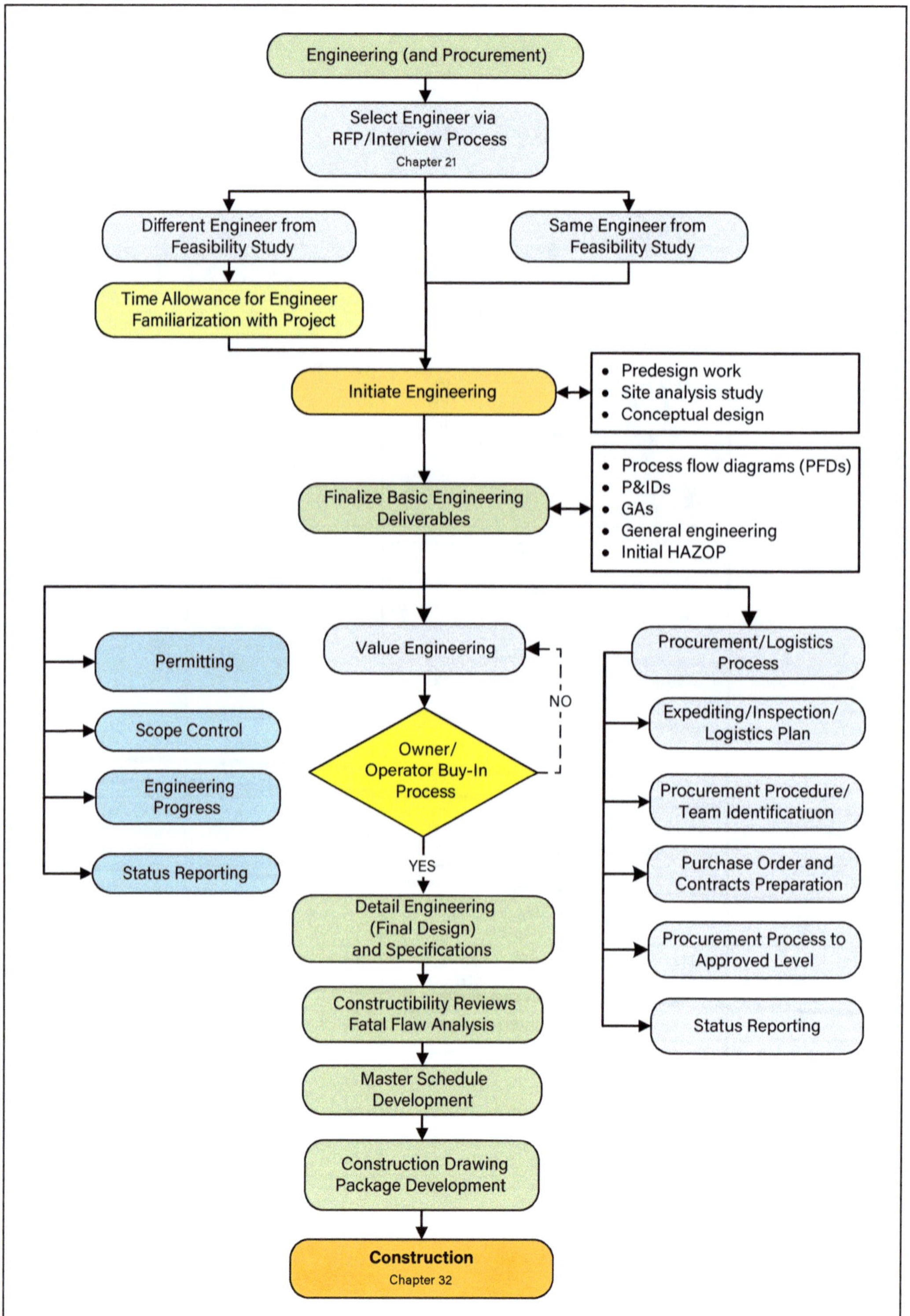

Figure 16.2: Engineering

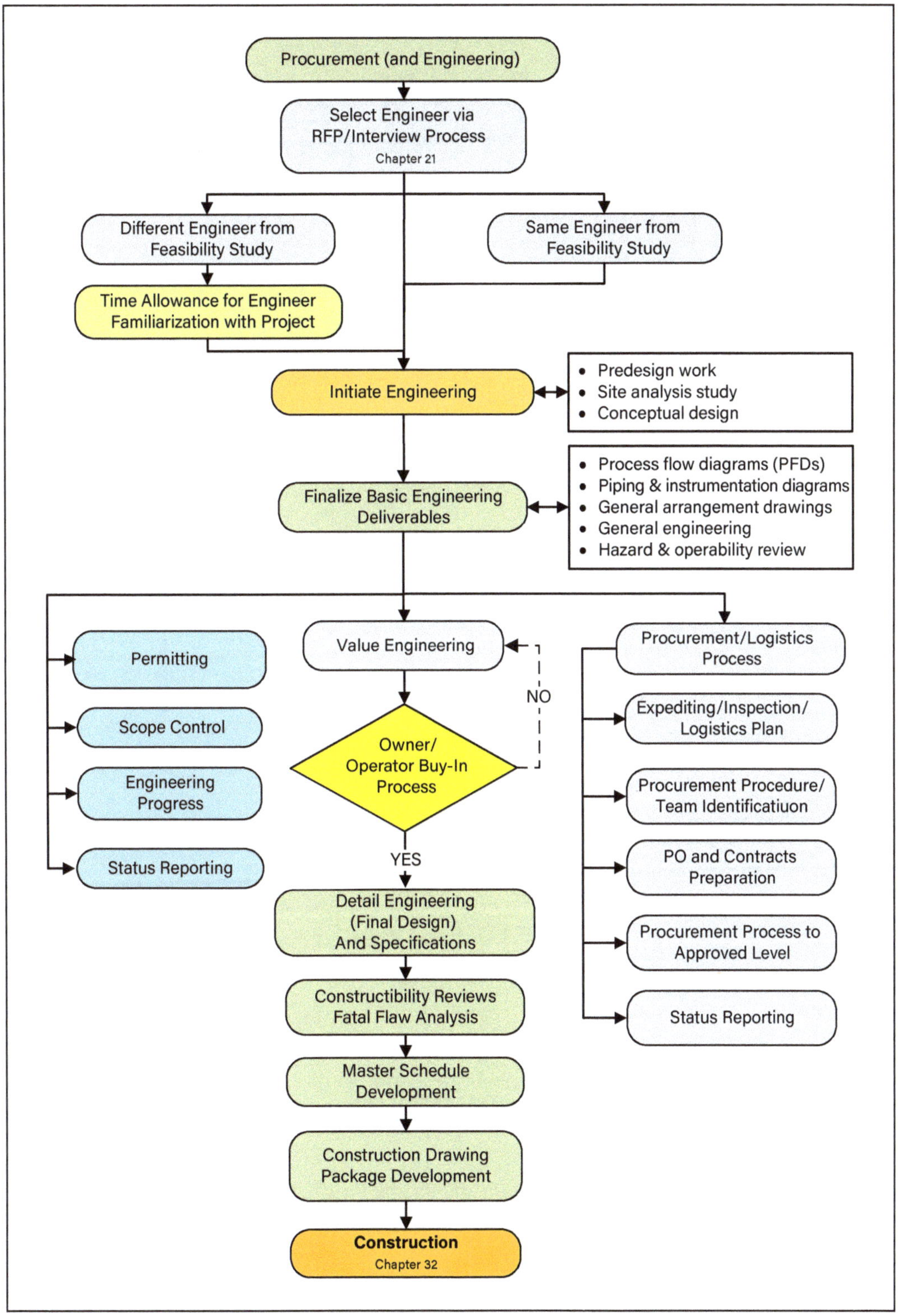

Figure 16.3: Procurement

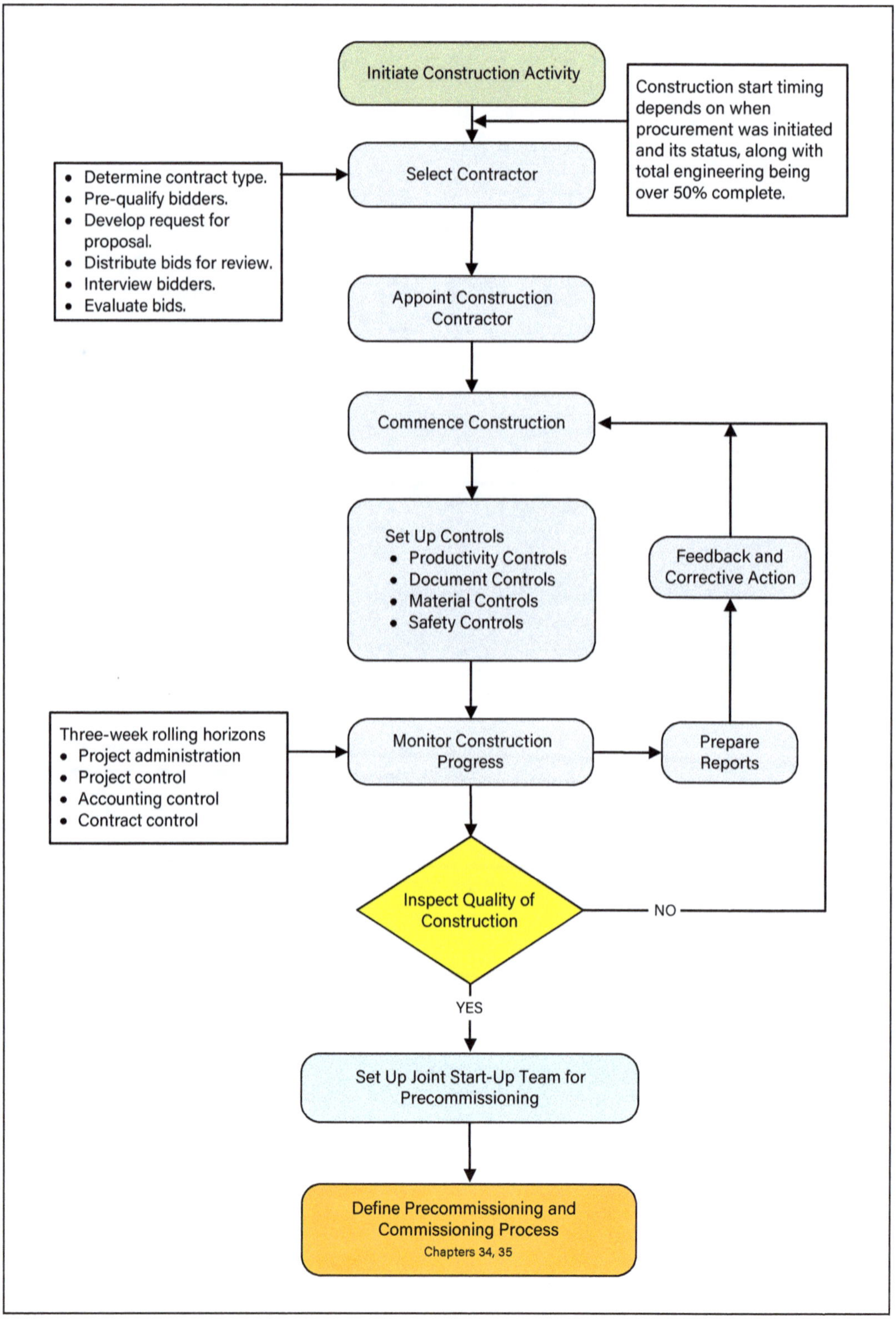

Figure 16.4: Construction

ENGINEERING

No	Item	Status	Date
1	Engineering (EPCM?) contractor—appointed		
2	Project scope of work—set		
3	Project coordination master schedule—established		
4	Drawing identification protocol—adopted		
5	Predesign phase—undertaken		
	A. Engineering basis—prepared		
	B. Engineering work program—set in conjunction with Owner		
	CADD system, software, and trained personnel—in place		
	Progress measurement procedures—defined		
	Engineering master schedule—established		
	Labor-hour forecast—issued		
	Scope change approval process—established		
	Forecasting and trending—instituted		
	C. Engineering procedures—set up		
	D. Program administration and coordination protocols—established		
	E. Target percentage of engineering for each project phase—set		
	F. Progress and status reporting requirements—defined		
6	Site analysis study—complete		
	A. Site information—gathered		
7	Conceptual design—developed		
	A. Schematic conceptual designs—issued		
	B. Probable cost estimate—derived		
	C. Owner authorization to proceed to basic engineering—approved		
8	Basic engineering—conducted		
	A. Functional philosophy—established		
	B. Design development steps—undertaken		
	Applicable standards and codes—adopted		
	Design basis and design criteria—issued		
	PFDs—fixed		
	Initial equipment list—produced		
	C. Standardization of systems and components—set with Owner		
	D. Basic engineering drawings, calculations, and specs—issued		
	E. Compliance verification—obtained from regulatory agencies		
	F. Owner authorization to proceed to detail engineering—approved		
9	Detail engineering—completed		
	A. Design freeze—agreed with Owner		
	B. Final value engineering trade-off studies—completed		
	C. Environmental impact, permits, social considerations—imposed		
	D. Vendor review comments—incorporated		
	E. Control philosophy (expert system, PLC and/or DCS)—set		
	F. Operations buy-in—accomplished		
	G. Constructability review—undertaken		
	H. HAZOP and fatal flaw analysis—completed		
	I. In-house drawing reviews and checks—complete and signed off		

ENGINEERING (continued)

No	Item	Status	Date
10	Engineering deliverables—issued		
	A. Design criteria (engineering and process), standards, and codes		
	B. PFDs, with mass balances		
	C. P&IDs		
	D. Engineering specifications		
	E. Specialty specifications criteria and qualifications		
	Quality		
	Environmental		
	F. Calculations		
	G. Value engineering trade-off studies		
	H. Control philosophy (expert system, PLC, and/or DCS)		
	I. Data sheets		
	J. Equipment and pipeline lists		
	Mechanical equipment		
	Valves and pipelines		
	Electric equipment, motors, and instruments		
	K. Cable and conduit layouts		
	L. Electrical and instrumentation schematics		
	M. O&M manuals and training protocols		
	N. Plot plan and GAs		
	O. Detail engineering drawings (Rev 0)—issued for construction (IFC)		
	P. Test procedures and inspection plans		
	Q. Test results (materials, chemical and physical)		
	R. Regulatory filing requirements		
	S. Start-up plans		
	T. Purchase requisitions—issued to procurement group		
	U. Material control procedures		
	V. Records retention instructions—issued to documents group		
11	Specification deliverables—issued		
12	File completed drawings and specifications with authorities—delivered		
13	Cost-estimating input—provided		
14	Construction bidding support—provided		
	A. Construction packages (drawings, specifications, etc.)—prepared		
	B. Bid evaluations—reviewed for compliance		
	C. Substitutions—evaluated		
15	Field engineering role and duties—defined		
	A. Design conformance—monitoring and reporting		
	B. External testing and inspection reports—oversight and approval		
	C. Shop drawings—review and approval		
	D. Latest issue control drawing list—complete set maintained		
	E. Vendor drawings and manuals—obtained in a timely way		
	F. Field orders—prepared for the project manager		
	G. Construction quality and site reports—maintained on file		
16	Project closeout duties—undertaken		
	Record (as-built drawings)—completed and delivered to Owner		

PROCUREMENT

No.	Item	Status	Date
1	Purchasing provider (EPCM contractor?)—appointed		
2	Procurement manager and procurement team—in place		
3	Procurement strategy—developed		
	A. Procurement roadmap—created		
	B. Categorization of goods to be procured—undertaken		
	Critical (high-impact and high-value) items—identified		
	Long-lead-time items—identified		
	C. Supplier dynamics and market analysis—conducted		
	D. Standardization of parts and systems—agreed with Owner		
	E. Work package segmentation—accomplished		
	F. Total value versus total cost package drivers—understood		
	G. Global perspective and local supplier support—adopted		
	H. Negotiations script—prepared		
	I. Alignment of EPCM and Owner procurement—established		
	J. If an OEM chosen as deliverer of part of SOW—roles understood		
4	Procurement responsibilities (Owner vs. EPCM vs. OEM)—defined		
5	Engineering support process and status—reviewed and understood		
6	Procurement plan—set in conjunction with Owner		
	A. Procurement procedures—established		
	Progress measurement (and S curves)—defined		
	Expediting program and schedule—established		
	Labor-hour forecast—issued		
	B. Progress and status reporting requirements—defined		
	Forecasting and trending—instituted		
	Scope change approval process—controlled by project manager		
	Cost and cash flow reports—delivered to project controls		
7	Procurement documents—prepared		
	A. General terms of purchase—complete		
	B. Engineer specs and material data sheets (MRQs and MRPs)—complete		
	C. Equipment lists—final		
	D. International sourcing policies—in hand		
	E. POs		
	Warranty and guarantee requirements—determined		
	Vendor assistance at start-up—defined		
	M anuals publication support—outlined		
	Spare parts policy—issued		
	F. Construction contracts		
	Insurance and bonding requirements—determined		
	G. Issues resolution and adjudication—addressed		

PROCUREMENT (continued)

No.	Item	Status	Date
8	Sourcing options—determined		
	A. Owner's, EPCM's, and OEM's preferred vendor lists—consulted		
	B. Owner's, EPCM's, and OEM's supplier alliances—utilized		
	C. Bidders list—developed		
	D. Requests for quotation (RFQs)—solicited		
9	Approved supplier list—prepared		
10	RFPs—issued specifying bidder deliverables		
11	Bid evaluations—completed		
12	Bidder selection—identified and agreed to by Owner		
	A. PO business terms (warranties, terms)—negotiated		
	B. Supply risks—assigned to natural owner of each risk		
	C. Vendor drawings delivery date—set		
13	PO—awarded, if negotiations successful		
14	Expediting and inspection:		
	A. Expediting program and schedule—established		
	B. Expeditors—appointed		
	C. Shop surveillance and quality assurance plan—conducted		
15	Logistics:		
	A. Logistics survey—undertaken		
	B. Logistics plan—developed		
	C. Bar-coding system for materials tracking—adopted		
	D. Freight forwarder and receiving port transit agent—hired		
	E. Materials tracking program—installed		
16	Site materials management:		
	A. Warehouses and laydown yards—established		
	B. Quality conformance—inspections conducted		
	C. Goods receipt in sync with procurement paperwork—OK		
	PO register—maintained		
	D. Vendor data log—maintained		

CONSTRUCTION

No.	Item	Status	Date
1	SOW for construction stage—set		
2	Engineering or EPCM contractor—previously engaged		
	A. Greater than 50% of engineering drawings at IFC status—OK		
	B. Procurement of critical materials meets construction need—OK		
3	Construction strategy—fully developed		
	A. SOW broken into logical, manageable-size packages—done		
	B. Packages reflect how construction is to be controlled—OK		
4	Construction responsibilities—defined		
	A. Contracts administration—appointed		
5	Selection of construction contractor (if not EPCM)—undertaken (refer to Checklist 21.1)		
	A. Preferred contract type for SOW—determined		
	B. Prequalification of construction firms—met via RFQs		
	Experience, capability, availability, and location—OK		
	Compatibility, financial stability, and commitment—OK		
	Flexibility with respect to contractual Issues—OK		
	C. RFP—developed		
	D. RFP—distributed		
	Bidding instructions (specifying bidder deliverables)		
	SOW (including project schedule)		
	Project description and specifications		
	Form of agreement		
	Bonding and insurance requirements		
	E. Interview of bidders on project site—undertaken		
	F. Bid assessment: evaluation ranking form—completed		
	G. Winning contractor(s)—selected with Owner		
	Contract agreement—negotiated		
	Appropriateness of contract type—reviewed		
	Basis of compensation—established		
	Contract incentives and penalties—considered		
6	Construction contractor(s)—appointed, if negotiations are successful		
	A. Insurance certificates—provided by contractor(s)		
7	Preconstruction meeting with site contractors—conducted by project manager		

CONSTRUCTION (continued)

No.	Item	Status	Date
8	Construction plan—finalized		
	A. Possible modification of PEP—ascertained		
	B. Subcontractor management and work apportionment—fixed		
	C. Work rules and codes of conduct—established		
	D. Project goals—reiterated		
	E. Ultimate construction workforce size—determined		
	F. Level 4 Execution Schedule—issued for all contractors		
	G. Multiple Level 5 Detailed Short-Term Schedules—prepared		
	H. Quality compliance plan—in hand and being followed		
	I. Risk management program—understood		
	J. Off-site preassembly, modularization, and fabrication—maximized		
	K. Advance site preparation work—undertaken		
	L. Temporary facilities—provided		
	M. Maximum number of work fronts—developed		
	N. Work-site services and security—arranged		
9	Reporting program to monitor and control work progress—established		
	A. Physical progress vs. plan using EVM		
	B. Labor complement and productivity monitoring vs. plan		
	C. Cost monitoring		
	D. Schedule adherence		
	E. Change management and control		
10	Material management control process—established		
	A. Site delivery times—continuously tracked		
11	Proper safety and health procedures—in place		
12	Requisite permits—obtained		
13	Environmental controls—in place		
14	Social acceptance and sustainability considerations—respected		
15	Document management and document control—established		
16	Preparation for precommissioning and commissioning—undertaken		
17	Construction completion—notifications prepared		
18	O&M manuals—all in place		
19	Procedures and team for operations start-up interface—established		
20	Punch list—produced, but clearance tied to precommissioning		
21	Mechanical completion—achieved by the constructor, but certification requires successful precommissioning verification for each system entity		
22	Construction contractual completion tests—acceptance certificate not issued by project manager until after successful commissioning		

SECTION 17

Project Completion

OBJECTIVE

There are two distinct steps to project completion: construction completion, followed by commissioning (including start-up steps and, ultimately, turnover of the project to the operations group).

Each section of these guidelines is designed to direct the reader to the corresponding chapter in *Project Management for Mining* (*PMfM*), which must be thoroughly understood for successful implementation. *PMfM* must be used alongside these guidelines.

REFERENCES

Item	*PMfM*	Page
Construction Completion	Chapter 33	627
Precommissioning	Chapter 34	637
Commissioning, Start-Up, Ramp-Up	Chapter 35	653
Turnover and Closeout	Chapter 36	669

CG Figure	Figure Caption	*PMfM* Reference
Figure 17.1	Construction completion	Figure 33.1, p.629
Figure 17.2	Precommissioning	Flowchart 34.1, p.651
Figure 17.3	Commissioning, start-up, ramp-up	Flowchart 35.1, p.668
Figure 17.4	Turnover and closeout	Flowchart 36.1, p.677

Checklist Title	*PMfM* Reference
Construction completion	Checklist 33.1, p.636
Precommissioning	Checklist 34.1, p.649
Commissioning	Checklist 35.1, p.666
Turnover and closeout	Checklist 36.1, p 676

SCOPE

Figure 17.1 shows how the project moves from a mass construction activity into the final phases leading to an operating facility. It is a critical chart for the Owner and EPCM to review early in the project and to agree on the terms laid out in the chart in order to assure effective completion of the project.

Construction completion occurs when:

- All systems composing the project have been built in accordance with:
  - Contract Drawings
  - Specifications and Manuals
  - Authorized Changes
  - Applicable Codes and Standards
  - Permit Restrictions
- All equipment has been correctly lubricated, aligned, and deemed safe and ready for precommissioning, including:
  - Mechanical
  - Electrical
  - Instrumentation

Mechanical completion means that an individual system, subsystem, or single piece of equipment within a project facility is complete enough for that unit system alone to mechanically, electrically, pneumatically, and hydraulically start (as the case may be) and perform a test run.

Precommissioning is testing, verification, and documentation showing that project installations are contract compliant. This step follows mechanical completion and shifts the focus from construction to commissioning and start-up of the project. While some construction work still remains to be completed, starting the plant now becomes the primary aim.

Commissioning follows completion of precommissioning. Commissioning involves running and testing all areas of the plant systems individually, then together with ore and process fluids (i.e., wet), to ensure the integrated systems can perform as anticipated and to correct any remaining problems before start-up.

Commissioning is under the control of the Owner. Care, custody, and control are transferred to the Owner from the construction contractor at the end of the precommissioning stage. The Owner is now legally responsible for both the new facilities and the operations and maintenance activities that take place during commissioning.

Start-up is the beginning of actual plant operations. This is a critical step covering start-up activities and then ramp-up of the plant to full design operational throughput or output. The plant runs continuously under load with material and/or solution after start-up, delivering mineral products (e.g., coal, copper, gold, molybdenum).

PROJECT CLOSEOUT

Project turnover activity must be complete before project closeout can be achieved. Some demobilization elements are initiated during commissioning, but most turnover (and closeout) activities take place while ramp-up is ongoing and are generally complete long before steady-state operations are achieved in the project facility. Full project turnover to the Owner's operators is achieved when the following conditions have been met:

- The facility is substantially ready for its intended use, with all key components operable.
- The Owner confirms that EPCM site responsibilities have ceased and that the work scope is satisfactorily complete
- Start-up is complete (though ramp-up can still be in progress)
- Performance tests are complete
- The final punch list has been developed, worked off, and accepted by the Owner
- Piping and instrumentation diagrams (P&IDs), as-built drawings, and revisions are final and delivered to the Owner
- All inspection, balancing certification, and test records are delivered to the Owner
- A list of necessary operational spare parts is provided to the Owner
- Operating and maintenance manuals are delivered to the Owner
- Equipment warranties and guarantees are documented and formally handed over to the Owner
- Construction inventory has been handed over to the Owner
- The facility is clean and free of construction equipment and debris
- The EPCM construction effort is completed and closed out. All site issues between the constructors and the Owner have been resolved.
- Determination of incentives or penalties (to either pay or collect) is complete
- Settlement of all "gray areas" in work scope has been reached with the Owner
- Each entity has been individually notified of commencement of warranty periods
- All payments have been made, including settlement of claims, incentives, and penalties. All liens should be removed at this time.
- Project accounting closure has been completed to the satisfaction of the Owner's asset-register manager
- Exit interviews of key staff have been conducted and lessons learned captured
- Personnel have all been reassigned and no one remains on project payroll
- The project AFE has been closed out and can no longer be charged against
- A Work Stop order (with date) has been issued by the PM to the team and to contractors
- Achievement of project goals has been assessed, and AFE results, including under and overspends, have been recorded.

Closeout is one of the toughest steps for a project to accomplish.

- Projects frequently go adrift and exceed both schedule and budget when project life enters its final 10%
- Causes of this undesirable project drag include:
  - Excess perfection syndrome and the craving to continually improve the facility
  - Burnout and loss of focus by the team as activity and intensity level wanes
  - Detrimental involvement in another project by team members
  - Lack of another project for project personnel to go to

Lessons Learned

The final project closeout step includes a candid look back at the entire project.

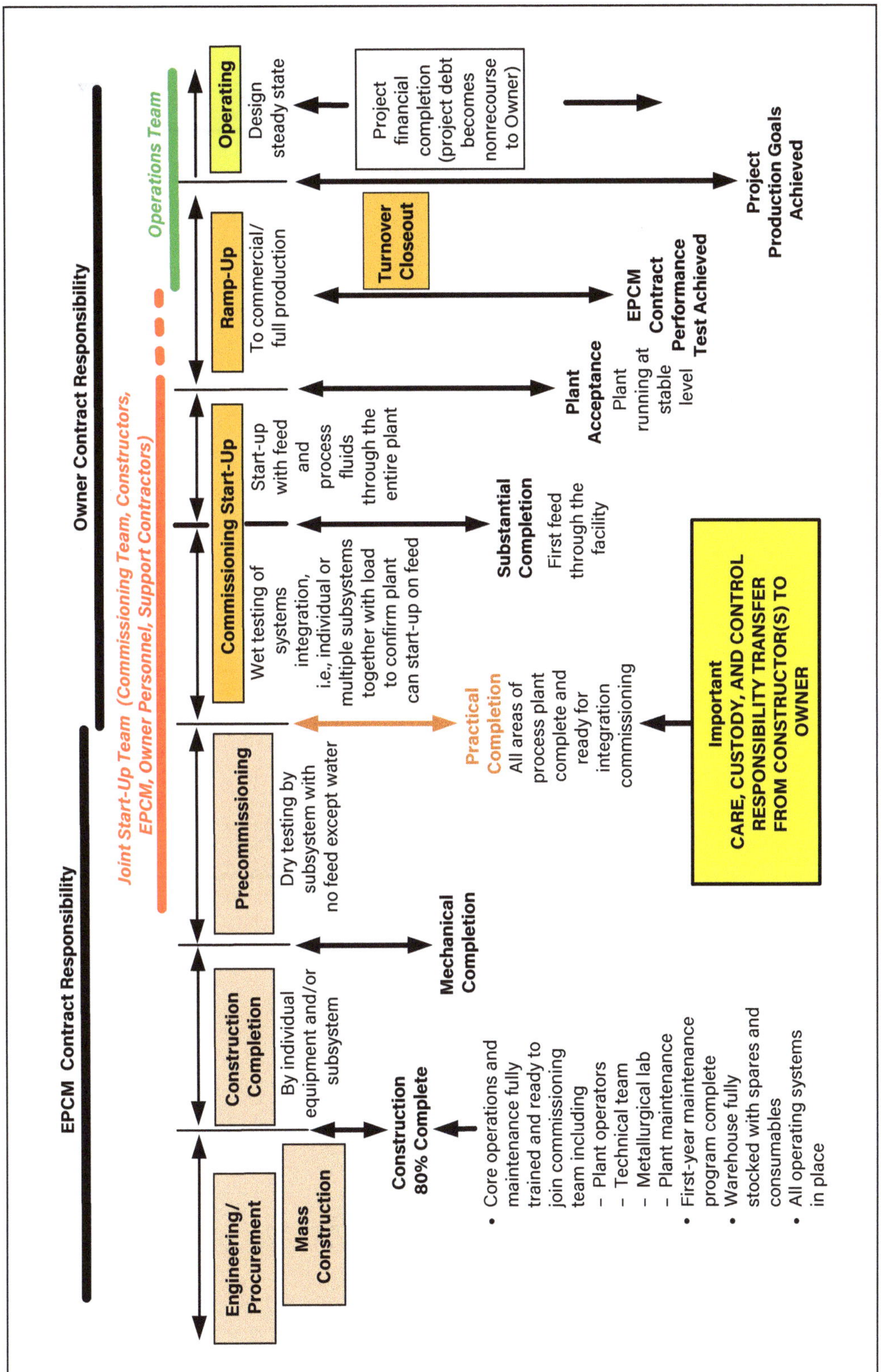

Figure 17.1: Project completion

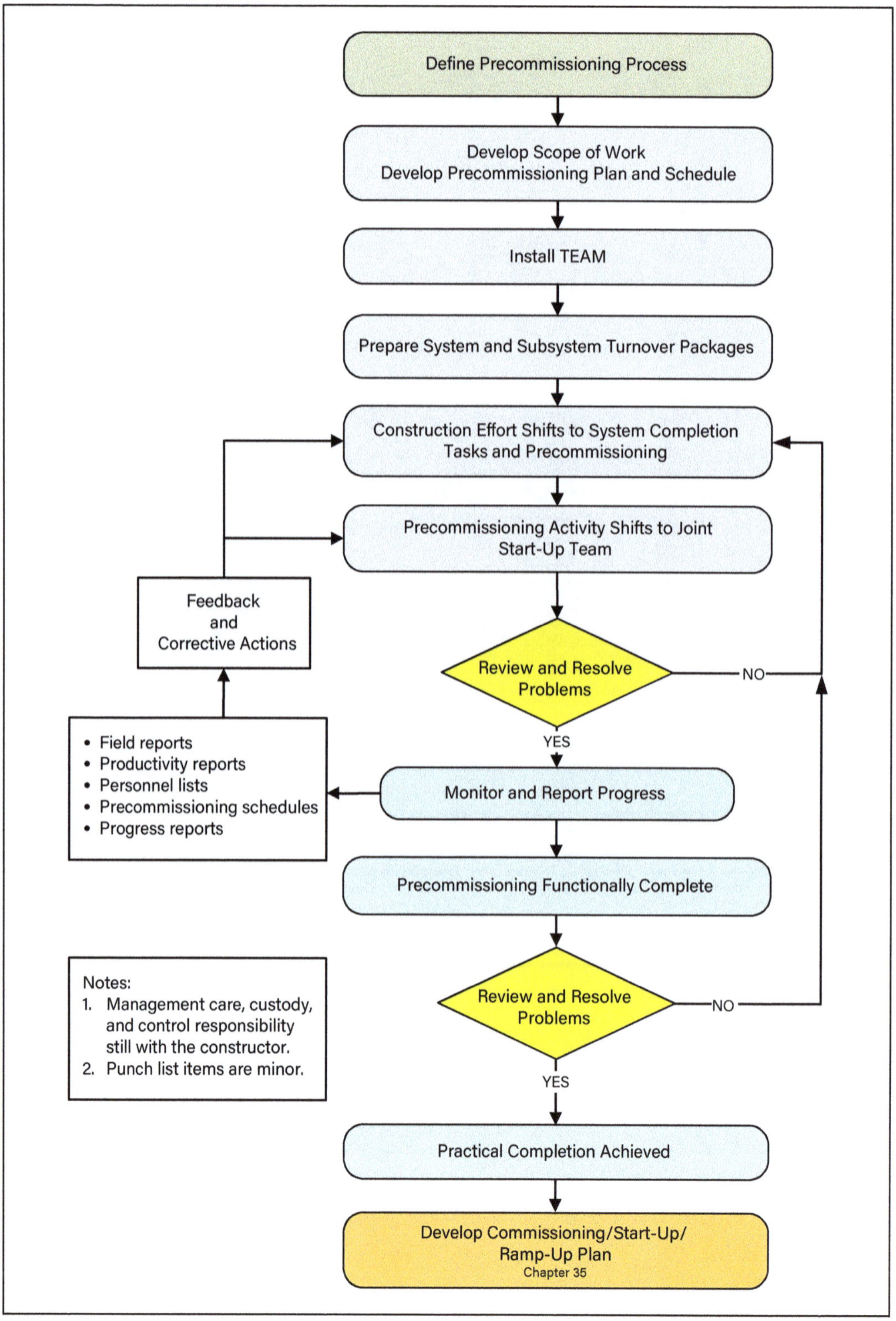

Figure 17.2: Precommissiong

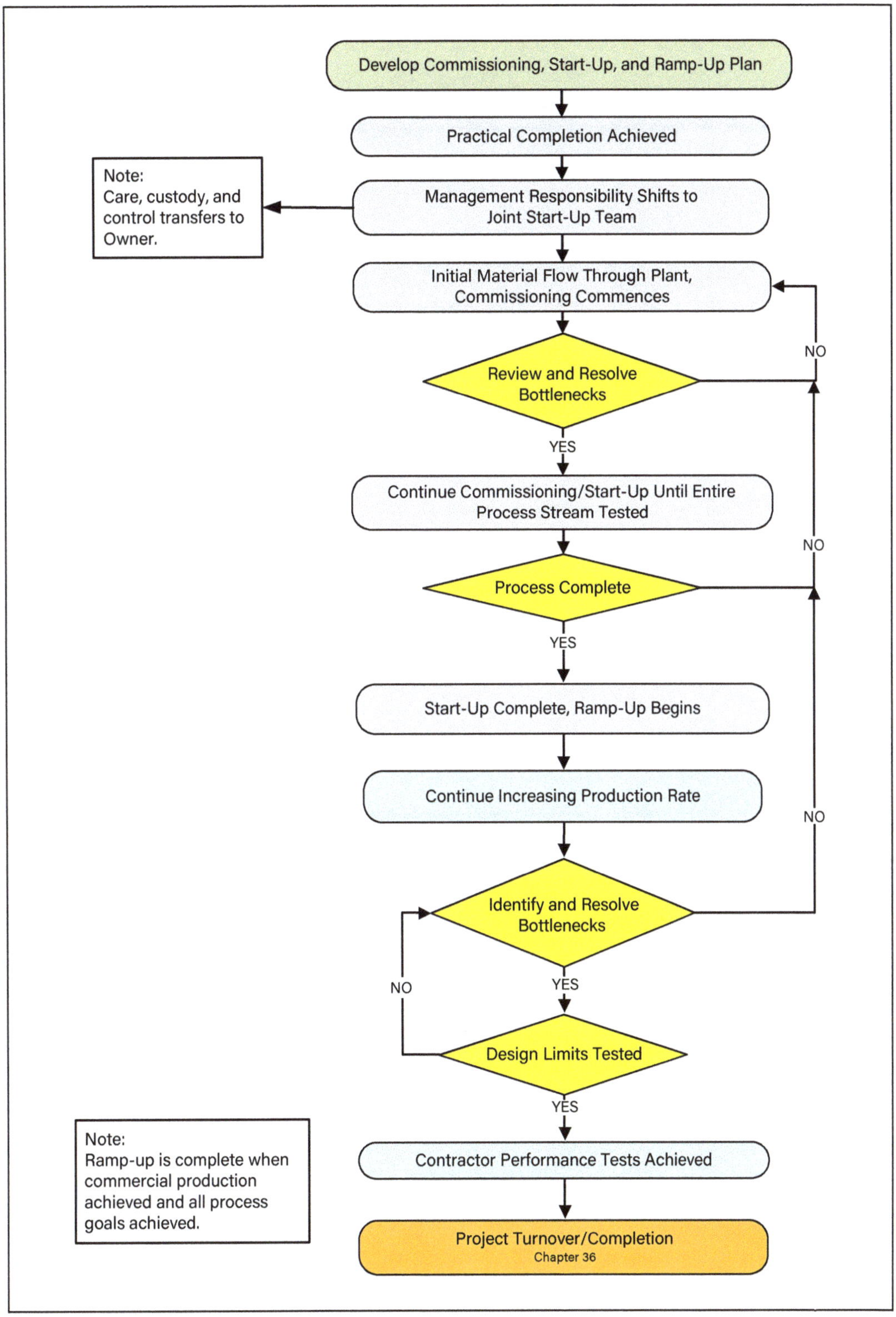

Figure 17.3: Commissioning, start-up, and ramp-up

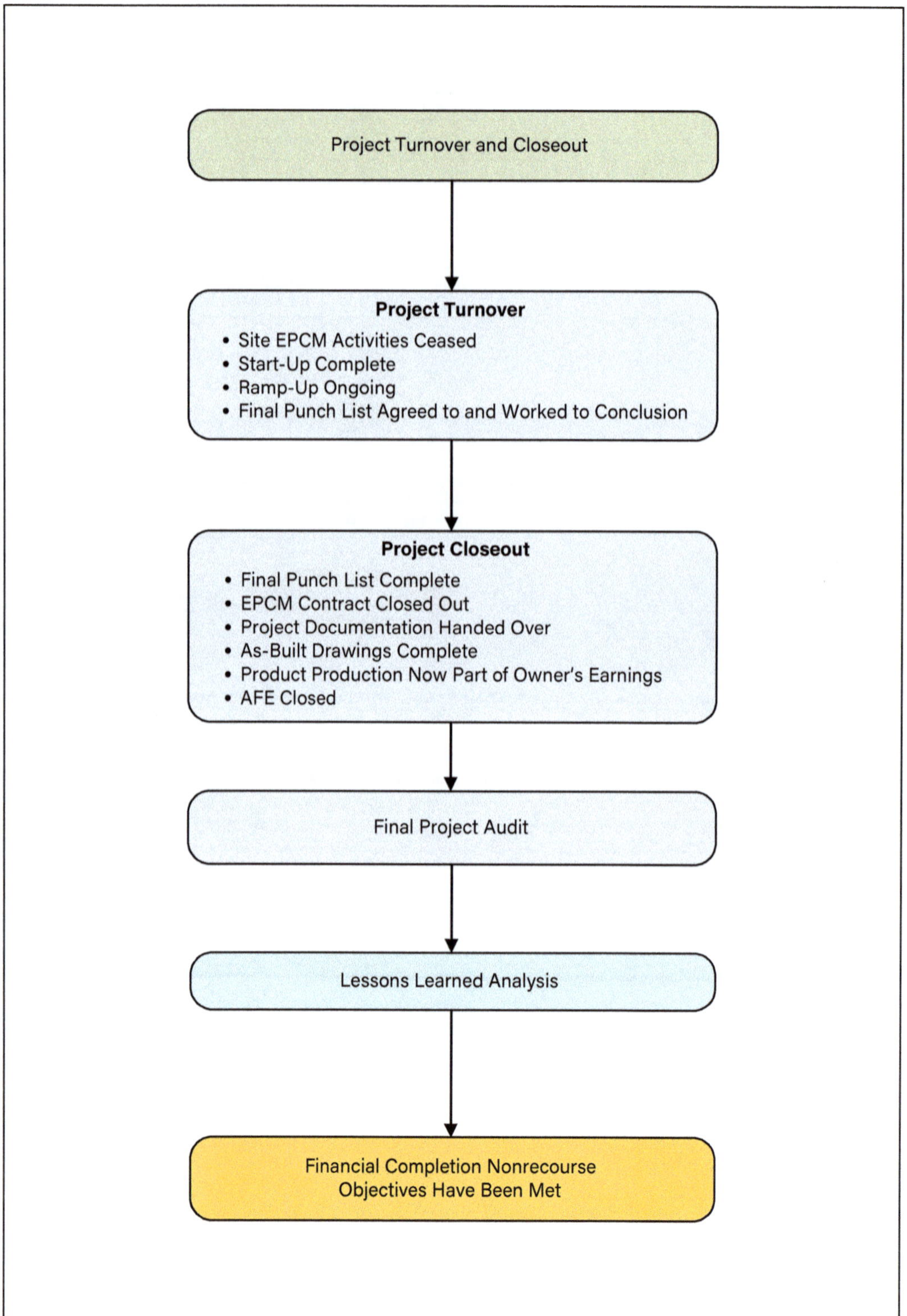

Figure 17.4: Turnover and closeout

CONSTRUCTION COMPLETION

No.	Item	Status	Date
1	Winning contractor(s)—selected with Owner		
2	Scope of work (SOW) for construction stage—set		
3	Construction plan—finalized		
4	Construction responsibilities—defined		
5	Contract agreement—negotiated		
	A. Appropriateness of contract type—agreed to		
	B. Precise, unambiguous definition of construction completion—set		
	C. Basis of compensation—established		
	D. Contract incentives and penalties—final disposition agreed to		
6	Insurance certificates—provided by contractor(s)		
7	Preparation for precommissioning and commissioning—undertaken		
8	O&M manuals—all in place		
9	Procedures and team for operations start-up interface—established		
10	Punch list—produced, but clearance tied to precommissioning		
11	Mechanical completion—certification achievement by constructor requires successful precommissioning verification for each system entity		
12	Care, custody, and control assumed by Owner—achieved upon the project's successful precommissioning		
13	Substantial completion—certification achieved by contractor upon successful commissioning verification of entire facility		
14	Construction contractual completion—acceptance certificate to be issued by project manager upon the project's successful commissioning		

PREECOMMISSIONING

No.	Item	Status	Date
1	Commissioning plan, from construction end through precommissioning, commissioning, start-up, and ramp-up into operations—fully developed		
	A. Precommissioning scope of work—prepared		
	B. Quality assurance to ensure that operations output meets the project's design quality objectives—installed		
	C. Constructor focus to complete individual systems—achieved		
2	Critical Owner operations personnel—all hired and in place		
	A. Manuals (operations, maintenance, and training)—all compiled		
	B. Operations training (including simulation programs)—complete		
	C. Environmental awareness indoctrination as to applicable environmental laws and regulations, as well as to permit conditions and social obligations—complete		
3	Precommissioning team (the team)—assembled		
	A. Team leader (experienced commissioning expert)—hired		
	B. Commissioning specialists—engaged		
	C. Key construction personnel—assigned		
	D. Owner operators (future plant personnel)—hired and trained		
4	Precommissioning team functionality:		
	A. Organized under joint construction and operations control—OK		
	B. Reports to project manager during precommissioning—OK		
	C. Plant facility care, custody, and control contractually under control of construction contractor, who reports to project manager—OK		
5	Precommissioning plan—produced		
	A. Plant divided into logical systems and subsystems—complete		
	B. Entity-by-entity schedule of activities—developed		
	Scope of work for each individual entity—established		
	Checklists and log sheets for each entity—prepared		
	C. Milestones—set		
	D. Individual system turnover packages—prepared		
	E. Monitoring and reporting system—in place		
	Weekly summary reporting—established		
	Daily activity log—published		
6	Spare parts and requisite consumables—all purchased and in place		
7	Safety compliance of each individual system and subsystem—verified		
8	Mechanical completion—verified for each system and subsystem		
	A. Mechanical completion acceptance certificates—issued		

PREECOMMISSIONING (continued)

No.	Item	Status	Date
9	Precommissioning activities—conducted by the team		
	A. Systems isolation for precommissioning testing—established		
	B. Manual dry commissioning—conducted		
	C. Electronic dry commissioning—conducted		
	D. Systems checkout and exercising on water and air—undertaken		
	No-load systems check on local control—run		
	No-load systems check on remote control—run		
	E. Punch list—produced		
	F. Problem identification and resolution process—derived		
	Punch list Items—Reduced to minor, or de minimis		
10	Owner operations personnel trained during precommissioning process		
	A. Witnessing, observance, and assistance provided by team—OK		
11	Practical completion—achieved		
	A. Sufficient systems precommissioned to allow material to be processed through those systems		
	B. No conditions remaining that can compromise commissioning		
	C. Precommissioning team has ensured that the plant is ready for Owner operations to take care, custody, and control		
12	Plant ready for commissioning—ascertained		
13	Owner signs for care, custody, and control—done		

COMMISIONING

No.	Item	Status	Date
1	Commissioning plan, from construction end through precommissioning, commissioning, start-up, and ramp-up into operations—fully established		
	A. Commissioning and start-up scope of work—prepared		
	B. Manuals (operations, maintenance, and training)—all compiled		
2	Commissioning and start-up TEAM—still in place from precommissioning		
	A. Team leader and commissioning specialists—continue as is		
	B. Key construction personnel—held over		
	C. Owner operators (future plant personnel)—hired and trained		
	Operations personnel training—complete		
	D. Vendor representatives—added to TEAM		
3	Commissioning team functionality:		
	A. Organized under joint construction and operations control—OK		
	B. Reports to project manager during commissioning—OK		
	C. Plant facility under care, custody, and control of Owner—OK		
	D. Constructor retains responsibility for plant to function—OK		
4	Commissioning plan—produced		
	A. Scope of work for each stand-alone system—established		
	Checklists and log sheets for each system—prepared		
	B. Spare parts and requisite consumables—confirmed in place		
	C. Testing devices—provided, with suitable calibrations		
	D. Schedule—modified by precommissioning feedback		
	E. Methodology for introduction of raw product—developed		
	Process balance requirements—identified		
	F. Performance monitoring and reporting—in place		
5	Safety compliance of each individual system and subsystem—re-verified		
6	Commissioning activities—conducted by the TEAM		
	A. Each individual system's isolation from rest of facility—verified		
	B. Practical completion and functional ability—system confirmed		
	C. Wet commission each system with no load (i.e., air, water)		
	Safety and environmental systems—commissioned		
	Utility systems—commissioned		
	Alignment runs of materials-handling systems—performed		
	Controls, weightometers, etc.—calibrated		
	D. First fills of reagents, grinding media, etc., before ore introduction		
	E. Wet commission each system under load (ore and reagents)		
	Stable conditions for each system—achieved		
	Operational parameters (flow, velocity, rate, etc.)—verified		
	Operational design performance—consistently achieved		
	Tests to confirm design quality achievement—performed		
	Controls system programming and tuning—conducted		
	F. Performance data—gathered, analyzed, and reported		
	G. Problem identification and resolution process—in place		

COMMISIONING (continued)

No.	Item	Status	Date
7	Owner operations personnel trained during commissioning process:		
	A. Witness, tuning, and stop–start assistance provided to TEAM—OK		
8	Commissioning completion—achieved when each stand-alone system individually reaches its designated continuous safe operational performance using feedstock, with all controls functioning as per design		
9	Start-up activities—accomplished mostly by future operations personnel:		
	A. Support personnel (construction, EPCM, and vendors)—on hand		
	B. Sufficient systems individually successfully commissioned to allow processing of raw material through the full facility—OK		
	C. Raw material processed though entire facility—achieved		
10	Substantial completion achieved:		
	A. The full facility has successfully run continuously with material at design load delivering the project's design quality products.		
	All design parameters—met or exceeded		
	All bottlenecks—identified and corrected		
11	Plant acceptance—requires documented verification that project can produce the contractually agreed and specified quality products at the contractually specified rate for the contractually agreed-on minimum period:		
	A. Mechanical completion, practical completion, and/or substantial completion—all certified as achieved		
	B. Plant availability and productivity—at specified levels		
	C. Contractual performance tests and guarantees—all met		
12	Demobilization of EPCM and construction—follows acceptance by Owner		
13	Ramp-up—accomplished entirely by future operations personnel		
	A. Schedule modified by commissioning and start-up feedback		
	B. Ramp-up to full production continues until:		
	Commercial production is achieved		
	All process-related goals are achieved		

TURNOVER AND CLOSEOUT

No.	Item	Status	Date
1	Project turnover:		
	A. Formal acceptance by Owner for constructed plant—finalized		
	B. EPCM site responsibilities—concluded		
	C. Final punch list—worked off and accepted by project manager and Owner		
	D. Construction inventory—handed over to Owner		
	E. Equipment warranties—provided to Owner in a formal document		
	F. As-built drawings—final revisions delivered to Owner		
	G. Computer disks—labeled and purged of all but the latest data		
	H. Construction and commissioning efforts—complete		
2	Project management closeout:		
	A. Project site—clean and free of construction debris		
	B. All constructor contracts and supplier invoices—closed out		
	C. Final invoices and holdbacks—approved and closed out		
	D. All claims—settled		
	E. Amounts of incentives and penalties to pay or collect—determined		
	F. All Owner and EPCM personnel—off the project payroll		
	G. Re-assignment of all Owner project personnel—complete		
	H. Gray area items (in or out of scope)—finalized with operations		
	I. Environmental compliance—responsibility transferred to Owner		
	J. Project execution AFE—closed		
3	Project accounting closeout:		
	A. All invoices and holdbacks—paid		
	B. Project incentives and penalties—paid or collected		
	C. Project AFE—closed out by Owner's accounting department		
	D. Underage or overage to capital AFE budget—documented		
	E. Capital property, plant, and equipment ledger and asset register— handed over to Owner		
	F. Depreciation schedule—handed over to Owner		
4	Final project audit—undertaken		
	A. Achievement of project goals—assessed and documented		
5	Exit audit with key project personnel—conducted		
	A. Lessons learned—documented		
6	Project financial completion—determined		
	A. Financial completion tests—all achieved		
	B. Senior project debt—goes nonrecourse to Owner's company		
7	Grand opening ceremony—scheduled by Owner's CEO		
8	Production—now a part of Owner's operational earnings		
	A. Start-up complete, but ramp-up can be ongoing		

SECTION 18

Brownfield Project

OBJECTIVE

Project Management for Mining (PMfM) is the standard for mining projects, whether greenfield or brownfield. A project is generally considered brownfield whenever it is undertaken at an operating property. The size of brownfield projects can vary significantly (*PMfM* Chapter 38).

Whatever the driving force for a brownfield project, the instigator is nearly always the existing operator. This creates a different dynamic because, unlike greenfield projects, the project leader will likely be working for the Operations Manager. The reason is straightforward: in a brownfield situation, the operation is making money for the corporation, while the project is a drain on the corporate treasury. This is the opposite of a greenfield situation, where operations cannot make money until the project is successfully completed. The brownfield Project Manager must understand this different dynamic and purposefully collaborate with operations to achieve a successful outcome for both the project and the operation.

Each section of these guidelines is designed to direct the reader to the corresponding chapter in *PMfM*, which must be thoroughly understood for successful implementation. *PMfM* must be used alongside these guidelines.

REFERENCES

Item	*PMfM*	Page
Brownfield	Chapter 38	697

CG Figure	Figure Caption	*PMfM* Reference
Figure 18.1	Implementing feasibility study	New Figure
Figure 18.2	Opportunity to viability stage	New Figure
Figure 18.3	Project execution plan for brownfield	New Figure
Figure 18.4	Brownfield project setup and execution	Figure 38.2, p.703
Figure 18.5	Precommissioning, commissioning, and start-up	New Figure

Checklist Title	*PMfM* Reference
Brownfield	Checklist 38.1, p.706

SCOPE

Generally, for an expansion at a mature facility, the same level of effort may not be required in certain areas as for a new greenfield project. For example:

- Project Governance may largely fall under the site's existing Change Management procedures
- A risk management process is already in place
- Permitting, environmental management, and social and sustainability activities will tie into existing programs and controls
- Team assembly and organization may leverage existing site teams and engage earlier with operations
- Procurement and contract-agreement types may be smaller in scale and may leverage existing contractors and suppliers
- Operations readiness preparations may be limited to implementation of new technology or processes

Increased emphasis, on the other hand, is likely needed in the following areas:

- Opportunity evaluation focused on revenue improvement. Throughput or recovery improvements that affect revenue are likely to generate higher returns than cost reductions alone.
- Study of alternatives, with special focus on tie-in points to existing plants and processes
- Viability and planning stages, which may be more complex because of the current operation
- Risk mitigation, which must expand to include downtime impacts on the current operation
- Engineering of tie-ins and compatibility with existing facilities
- Controls at interface points and schedule alignment with the operation to ensure continuity of operations

Opportunity Stage

The opportunity stage is similar to that of a greenfield project. The scope or complexity may differ, but it should be pursued with full PFS or FS activities in mind.

Viability Stage Strategy

- Set battery limits around the existing operating property
- Determine who will deliver the FS
  - Define the deliverables from the FS
  - Develop a coordination plan among all parties, including Operations
- The output is a PEP (*CG* Section 12) or Project Charter (*CG* Section 3).

Viability Stage (*PMfM* Chapter 5)

Particular attention must be given to the transition path from scoping evaluation to FS.

- Ideas are typically identified by operators
- Operations often conduct a scoping evaluation but may not have the experience to include all components required for a proper study.
- It is necessary to guard against the tendency to underestimate project complexity, cost, schedule, and risk

The Viability Stage is the point at which the project development process begins.

- Viability Stage work should be conducted by an experienced project team.
- A Project Charter or PEP is prepared to guide the development process.
- Before starting the FS, conduct a gap analysis by a Project Delivery expert.

PEP for a Brownfield project

Brownfield PEPs require special consideration because they are developed at an existing operation. Developing a new major project in or around ongoing operations requires:

- Significant additional planning to ensure operations integrity.
- Coordination between operations and project teams.
- Resource -allocation agreements among all groups.
- Special procedures and controls for field execution of the project.
- Specific safety protocols related to the operating facility and the new construction.

Project Setup

Completion of project setup planning for execution requires:

- An experienced project team
- A completed FS
- A fully developed PEP
- Project procedures documented and distributed
- Operations-readiness plans in place
- A fully approved AFE supported by the above components and economic analysis

Execution requires:

- Implementation of an engineering plan
- Procurement and construction teams and plans ready to proceed
- Completion, commissioning, and closeout stages clearly defined
- Project controls, systems, and management in place

Operations Readiness

Brownfield projects must ensure a seamless process when bringing a new plant facility into an ongoing operation.

- Use checklists to conduct a formal review and confirm that the current operation is truly ready for the new processes to be introduced
  - Develop a strategy for operational capability
  - Implement an Operations Readiness Plan
  - Establish the appropriate organization and recruit or train personnel
  - Identify needs and install business systems
  - Prepare and implement policies, procedures, and SOPs
  - Undertake necessary training
  - Manage community and external stakeholders
  - Specify requirements for mine development
  - Match the needs of the process and support facilities
  - Align the project with existing operations

Special Considerations for Project Completion at an Operating Plant

- Ongoing operations can complicate any, some, or all project activities
- Activities from commissioning through start-up, and transfer of the facility to operations, must be very carefully planned and agreed to by all parties involved
- This critical step will make the difference between a successful start-up and a prolonged, difficult process

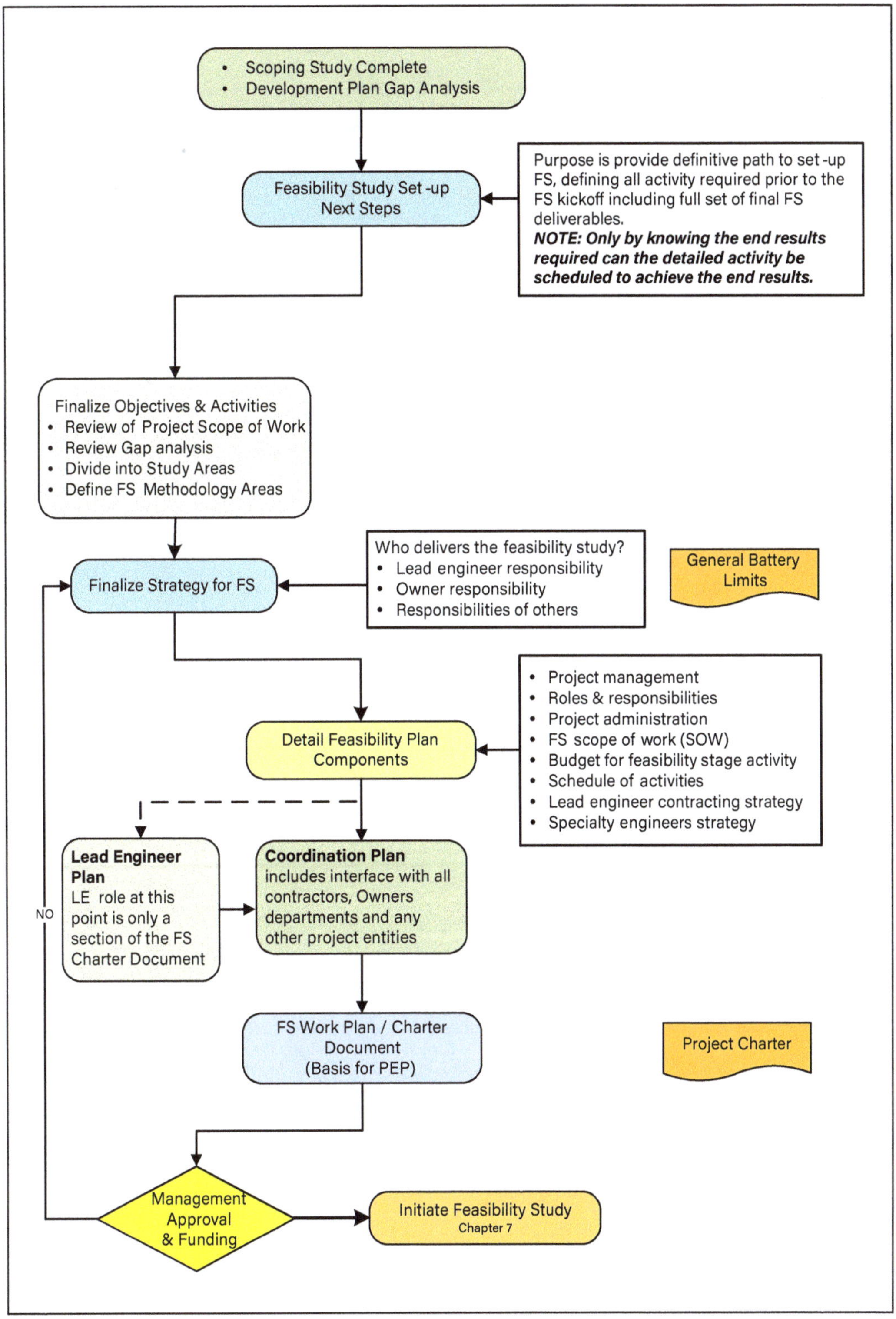

Figure 18.1: Implementing feasibility study

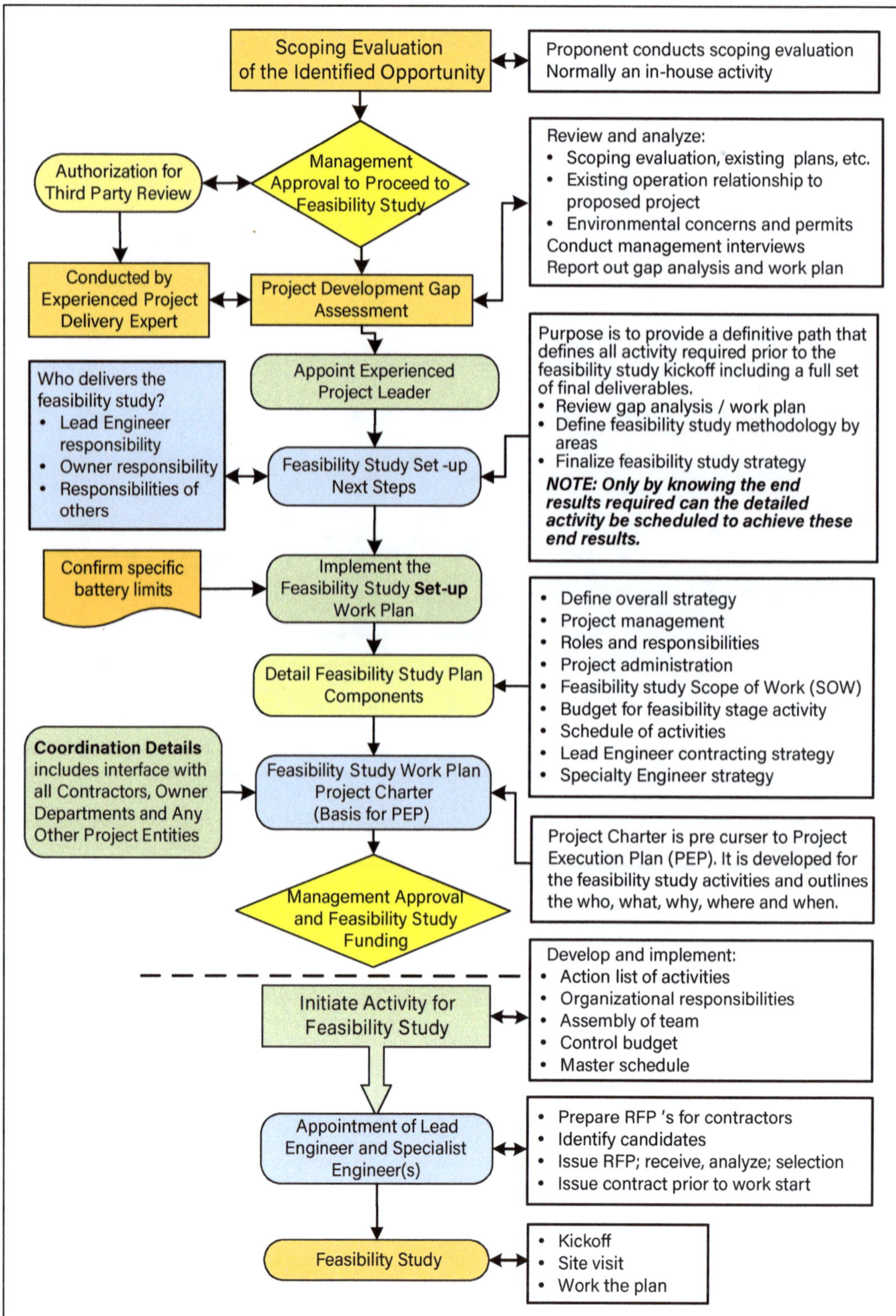

Figure 18.2: Opportunity to viability stage

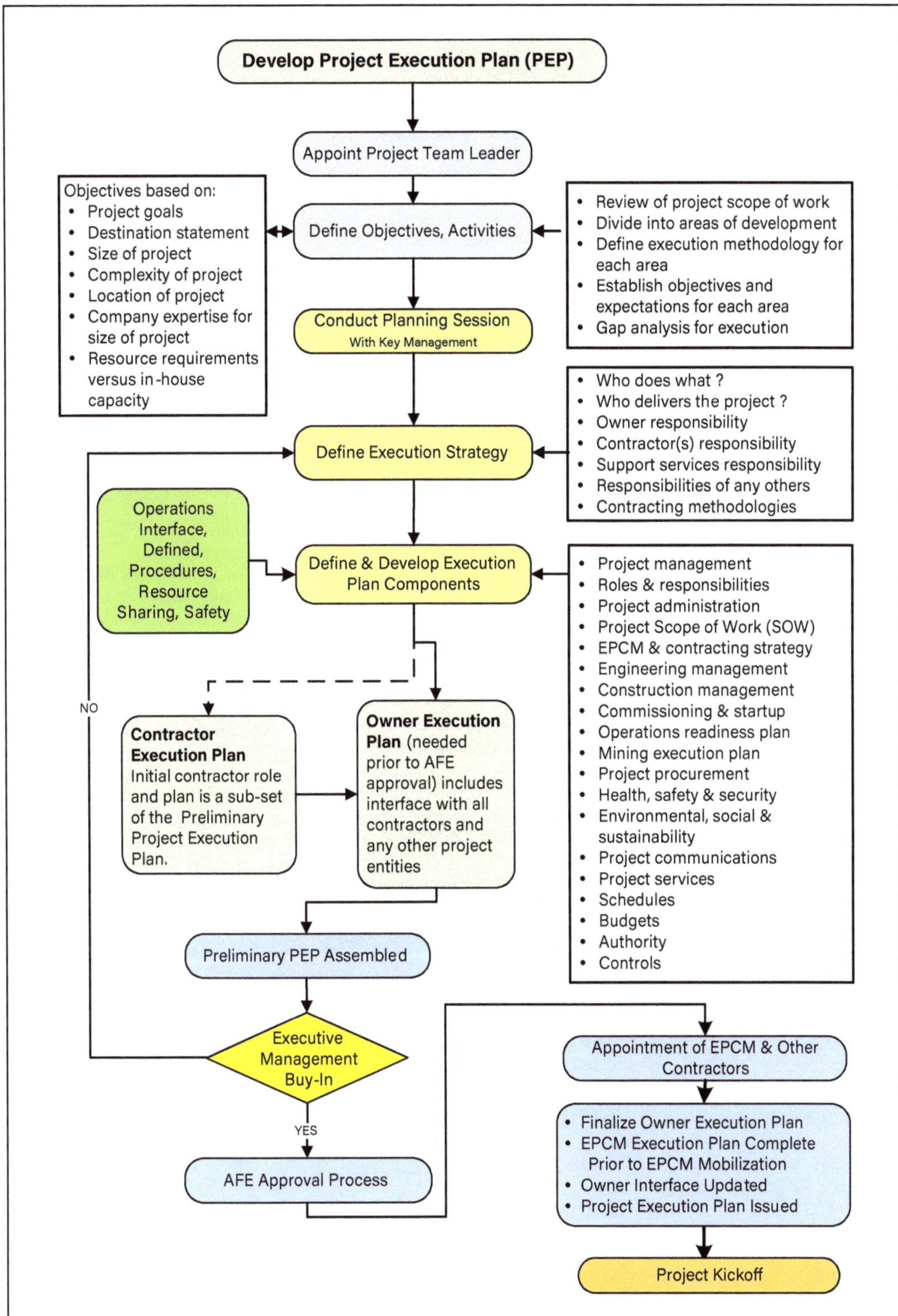

Figure 18.3: Project execution plan for brownfield

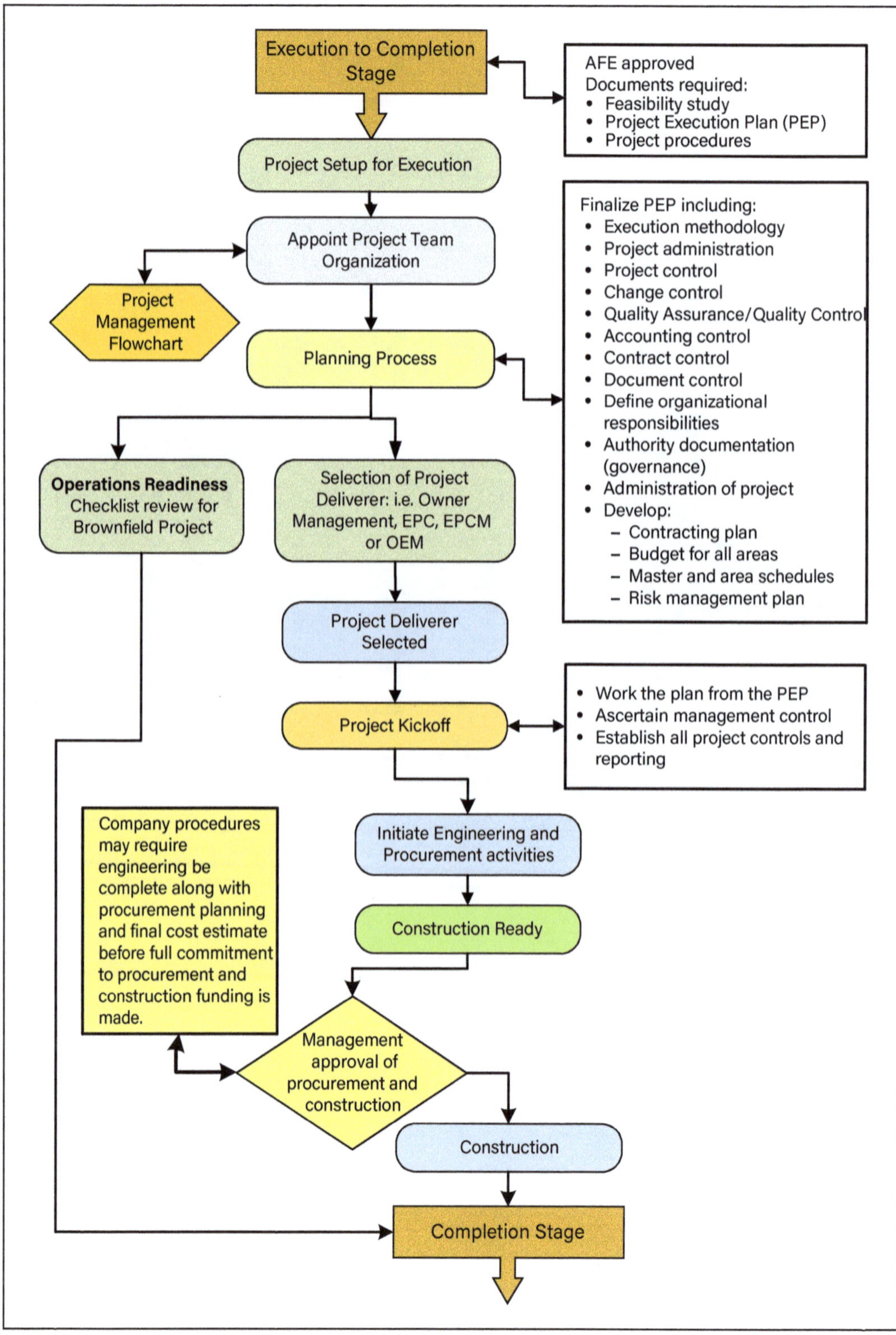

Figure 18.4: Brownfield project setup and execution

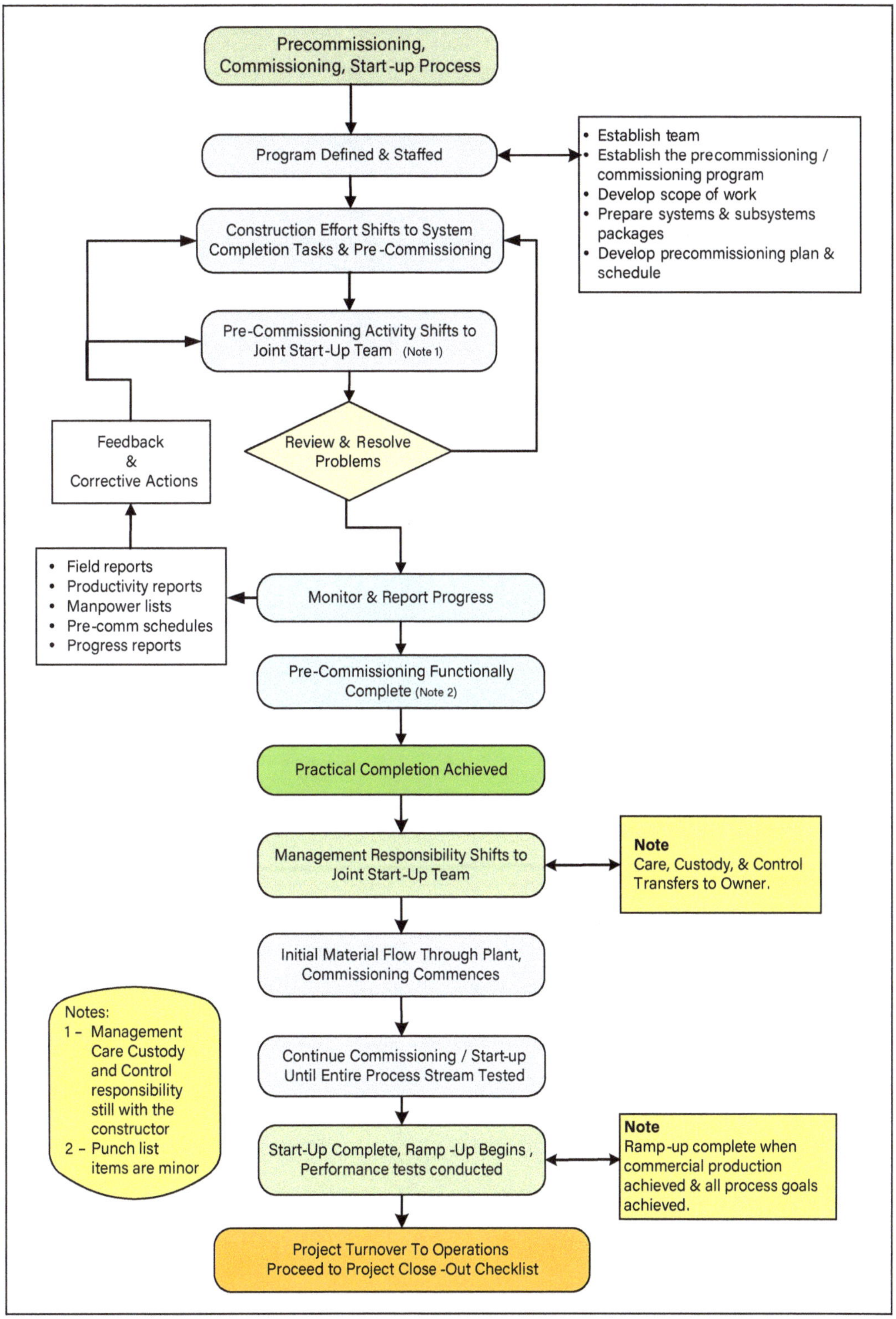

Figure 18.5: Precommissioning, commissioning, and start-up

BROWNFIELD PROJECT

No.	Item	Status	Date
1	Scoping evaluation correlated to the opportunity—complete		
2	Management approval to proceed, per *PMfM* Chapter 6—granted		
3	Project development gap assessment by a knowledgeable project delivery expert—conducted		
4	Experienced project manager—appointed		
5	Work plan for proceeding to the feasibility study—finalized		
	A. Development details based on gap assessment—produced		
6	Project battery limits—fixed with concurrence of operating group		
7	Project team—assembled		
8	Lead engineer for feasibility study (*PMfM* Chapter 21)—chosen		
9	Feasibility study, per *PMfM* Chapter 8		
	A. Preparation work—complete		
	B. Strategy and deliverables—finalized		
	C. Feasibility study plan components—detailed		
	D. Feasibility study—initiated		
	E. Project objectives and Owner's goals—fully aligned		
	F. Feasibility study—complete		
10	PEP, per *PMfM* Chapter 18—complete		
	A. Comprehensive risk mitigation plan—in place		
11	AFE for project execution, per *PMfM* Chapter 16—prepared		
	A. AFE guidelines—followed		
	B. AFE form—signed and submitted		
12	Management approval to proceed, per *PMfM* Chapter 17—received		
13	Setup of project for execution, per *PMfM* Chapter 19—satisfied		
14	Alignment with existing company safety policies—accomplished		
15	Reporting and information feedback system—established		
	A. Communications plan (internal and external)—implemented		
16	Follow-up on permitting and environmental issues—undertaken		
17	PEP and SOW—refined and frozen		
	A. Battery limits around existing operating property—set		
	B. Owner commitment on operator availability—confirmed		
	C. Operations readiness plan—developed		
18	Project deliverer selection, per *PMfM* Chapter 21—done		
19	Project procedures manual (PPM)—assembled		
20	Project controls, per *PMfM* Chapter 24—in place		
21	Pertinent social and sustainability issues—respected		
	A. Continuity of local community engagement—ensured		

BROWNFIELD PROJECT (continued)

No.	Item	Status	Date
22	Project kickoff, per *PMfM* Chapter 29—completed		
23	Final PEP—in place and being followed		
24	Engineering program, per *PMfM* Chapter 30—established		
25	Procurement plan, per *PMfM* Chapter 31—prepared		
26	Final approval for project execution, post estimate update—given		
27	Engineering—undertaken		
	A. Specialist consultants for execution support—hired		
28	Project procurement—functioning		
29	Engineering deliverables—furnished		
30	Construction program, per *PMfM* Chapter 32—set up		
	A. Quality of construction inspection protocols—created		
31	Field construction—executed		
32	Commissioning, start-up, and ramp-up—accomplished		
33	Project completion, per *PMfM* Chapter 33—achieved		
34	Turnover and project closeout, per *PMfM* Chapter 36—finished		

Glossary

Agreement-Based Decision Making/*Toma de Decisiones Basada en Acuerdos*: means that one person or role has the authority to make the final call, and others are expected to support and implement that decision once it is made—regardless of whether they personally agreed with it.

Authority/*Autoridad*: The right to apply project resources, expend funds, make decisions, and give approvals.

Authorization for Expenditure (AFE)/*Autorización del Gasto de Inversión:* The corporate approval document for capital projects and major expenditures. Generally, an AFE is only required for expenditures higher than a certain minimum level (e.g., $10,000) and/or an expected life of above a minimum number of years (typically set somewhere between 1 and 3 years). The form can also be used for soliciting, recording, and communicating approvals for expenditures among a company's management responsibility levels.

"Bankable" Feasibility Study/*Estudio de Factibilidad "Bancable"*: A study demonstrating project operational and economic viability that has sufficient engineering detail and a high enough accuracy that it can withstand rigorous third-party technical and financial institution review, to be capable of commanding nonrecourse external financing for funding of the requisite project capital.

Basis of Estimate (BOE)/*Bases del Estimado*: A clear, unambiguous statement of the underlying scope basis, pricing basis, methods, references, qualifications, assumptions, inclusions, and exclusions that were used to develop the cost estimate. The BOE describes how the estimate was derived and how the expected funding will be spent to achieve the given objective.

Basis of Schedule/*Bases del Cronograma*: A complete, clear, and unambiguous statement of the underlying principles, basis, facts, assumptions, references, qualifications, and exclusions that are required to develop the project schedule.

Battery Limit/*Límites de Batería*: A defined boundary around an area of responsibility that identifies and limits the scope of work for the system or facility as well as for the contractor charged with performing the work.

Brownfield/*(Not Translated)*: Projects within or around prior work that are constrained by that existing work.

Budget/*Presupuesto*: The planned and approved cost of the project.

Chapter/*Capítulo*: Refers to *Project Management for Mining* which is divided into chapters.

Commissioning/*Comisionamiento*: (Also known as wet commissioning) starts when enough components of a stand-alone system are mechanically complete and have been successfully precommissioned dry, such that practical completion has been reached for that individual system. Feed ore material and/or process fluids can now be introduced for the first time into the system.

Committee for Mineral Reserves International Reporting Standards (CRIRSCO)/*(Not Translated)*: The CRIRSCO reporting system aligns codes and standards from those national reporting organizations that are CRIRSCO members to the 2012 published CRIRSCO template. Adherence to this template provides assurance for companies and for the reviewing government agencies that the published mineral resources and mineral reserves have met the highest recognized professional standards. The most utilized CRIRSCO-compliant standards for resources and reserves are NI 43-1019 (Canada); JORC Code (Australia), SAMREC Code (South Africa), and PERC (Europe), but others also exist and are recognized (i.e., standards from Brazil, Chile, Colombia, India, Indonesia, Kazakhstan, Mongolia, Russia, and Turkey). A U.S. standard that meets CRIRSCO requirements, jointly drafted by the Society for Mining, Metallurgy & Exploration (SME) (a CRIRSCO member) and the National Mining Association, finally became effective on January 1, 2021, after two decades of internal wrangling and review.

Consensus Decision Making/*Toma de Decisiones por Consenso*: means that all parties must agree before a decision can proceed. Everyone at the table has effective veto power—the discussion continues until a solution is found that everyone can accept, or at minimum, that no one actively opposes.

Constructability/*Constructibilidad*: The optimum use of construction knowledge and experience in all project phases, from concept through execution.

Contingency/*(Not Translated)*: A specific provision added to a base estimate to cover indefinable items that have historically, by actual experience, been required but cannot be specifically identified in advance. These undefined elements are fully expected to occur in the forecast value of cost and/or schedule, but the exact nature and timing of their occurrence is indeterminate. It is not intended to cover scope changes or project exclusions.

Cost Reimbursable/*Costo Reembolsable*: A contract pricing method under which all allowable and reasonable costs incurred by a contractor in the performance of a contract are reimbursed in accordance with the terms of the contract. Such contract types include cost plus (time and materials), fee plus, and schedule of rates.

Critical Path/*Camino Crítico*: The sequence of interconnected project network activities (i.e., without any gaps) that add up to the longest overall duration path through the project from project start to project end. The critical path thus equates to the shortest time possible to complete the project and, as such, is a prime determinant of the project's duration.

Cutoff Grade (COG)/*Ley de Corte*: When calculating ore reserves or mining plans, the grade at which the mineral resource is no longer worthy of extraction is termed the cutoff grade. In most situations, this is the break-even grade, that is, the grade at which the mineral resource can no longer be processed at a profit, but this does not have to be so; it may be set higher or lower than the break-even grade depending on the particular goals of the resource owner.

Distributed Control System (DCS)/*Sistema de Control Distribuido*: A computerized system controlling the production lines of a process (or any kind of dynamic system) in which the controller elements are not central in location but are distributed throughout the system, with each component subsystem regulated by one or more controllers.

Engineering, Procurement, and Construction (EPC)/*Ingeniería, Procura y Construcción*: The engineering, procurement, and construction function of work, most often applied to the contract scope delivered by engineering and construction contractors who use their own direct-hire personnel for execution of the construction work element.

Engineering, Procurement, and Construction Management (EPCM)/*Ingeniería, Procura y Gestion de Construcción*: The engineering, procurement, and construction management functions of work. The term is most often applied to the single-contractor entity employed by the Owner to provide the three core project functions of engineering and design, procurement of materials and equipment, and management of field construction, all on behalf of the Owner.

Enterprise Resource Planning (ERP)/*(Not Translated)*: A cross-functional, computerized system driven by an integrated suite of software modules that supports key internal data transfer processes within a company's operations. ERP systems can run on a variety of computer hardware and network configurations, typically employing a database as a repository for information.

Environmental Impact Assessment (EIA)/*Evaluación del Impacto Ambiental*: A first-order environmental review mandated by U.S. and other country regulatory authorities.

Equator Principles/*Principios de Ecuador*: A risk management framework promulgated in 2003 and adopted by financial institutions for determining, assessing, and managing environmental and social risk in project finance. They are primarily intended to provide a minimum due diligence standard for project decision-making. Financial institutions will not loan money to projects that are unable to comply with these standards. The fourth iteration of the Principles was published in 2019.

Estimate at Completion (EAC)/*Estimación al Finalizar:* An earned value management (EVM) term for the expected total cost of a schedule activity, work breakdown structure (WBS) component, or the defined scope of the total project at completion. The EAC may be extrapolated based on performance to date, in which case EAC = budget at completion/Cost Performance Index or estimated by the project team based on other factors, in which case, it is called a revised EAC.

Float /*Holgura:* Free float is the amount of time that a schedule activity can be delayed without postponing the early start date of any immediately following schedule activities. Total Float refers to how long an activity can be delayed without delaying the entire project completion date.

Forecast/*Proyección*: An estimate, projection, or prediction of conditions and events in the project's future based on information and knowledge available at the time of the forecast. A forecast, as defined in this handbook, refers to an adjustment of the original estimated budget and/or schedule. The forecast (i.e., restatement of original base budget or schedule) is based on the project's actual past performance, performance to date, actual or expected future commitments, improved knowledge of scope and design quantities, and/or expected future performance, including all known information that could impact the project in the future.

Gap Analysis/*Analisis de Brechas*: A formal determination and documentation of the difference between current performance capability and the desired future performance potential.

General Arrangement Drawing (GA)/*Plano de Disposición General*: A drawing that shows the general layout of the facilities being designed. A layout drawing, as opposed to a detail drawing.

General Contractor/*Contratista General*: The main (or primary) construction contractor responsible for the physical construction of the whole project facility (or the substantial portion thereof). The general contractor may or may not be authorized to hire subcontractors to deliver portions of the project services.

Greenfield/*(Not Translated)*: Project starting from the beginning without any prior work constraining the designwork.

Guaranteed Maximum Price (GMP)/*Precio Máximo Garantizado*: A cost-type contract where the contractor is compensated for actual costs incurred plus a fixed fee subject to a ceiling price, which the contractor guarantees not to exceed. The contractor is responsible for cost overruns, unless the GMP has been increased via a formal change order. Savings resulting from cost underruns are returned to the Owner.

Hazard and Operability (HAZOP) Review/ *Revisión de Peligrosidad y Operabilidad*: A structured and systematic examination of the planned project processes and facilities conducted at least once (preferably twice) during the engineering phase of the project. A HAZOP review is a qualitative technique based on guide words and is carried out by a multidisciplinary team (including all the engineering discipline leads and key Owner operations and maintenance personnel) within a set of structured meetings led by a specialist facilitator to identify any potentially hazardous operations or conditions in the finished facility.

Independent Project Analysis (IPA)/*(Not Translated)*: A benchmarking consultancy advisory organization founded in 1987 by Edward Merrow in Reston, Virginia, whose mission is to improve capital project effectiveness.

Internal Rate of Return (IRR)/*Tasa Interna de Retorno (TIR)*: A rate of return used in capital budgeting to measure and compare the profitability of investments. The IRR is also referred to as the DCF-ROI (discounted cash flow return on investment).

Issued For Construction (IFC)/*Emitido para Construcción:* Drawings that have reached Rev 0 status and are ready to be used by constructors in the field.

Key Performance Indicator (KPI)/*Indicador Clave de Rendimiento*: A measurement taken of a process or function. The KPI is used as an indicator of the health or effectiveness of that process or function.

Lump Sum/*Suma Alzada*: A specific type of fixed-price contract under which the buyer or purchaser (Owner or client) agrees to pay the seller, provider, or contractor a fixed total amount for completing work within set battery limits or delivering a well-defined product, generally without requiring a cost breakdown.

Management Reserve/*(Not Translated)***:** A monetary provision held by senior management (typically the corporate executive committee) for possible changes in the project scope or quality. Management reserve is an allowance over and above contingency and escalation that reflects project-specific risk (e.g., physical, social, and economic change that may impact the cost estimate after it was properly prepared using appropriate thoroughness and professional standards).

Management Review Submission/*Presentación para Revisión por Parte de la Dirección***:** The support documentation for an Authorization for Expenditure (AFE) form prepared in the study stages of the project outlining the scope and project parameters, and which ultimately forms the basis for the Project Execution Plan (PEP).

Master Schedule/*Cronograma Maestro***:** A summary-level project schedule that identifies the major deliverables, work breakdown structure (WBS) components, and key schedule milestones.

Material Takeoff (MTO)/*(Not Translated)***:** A term used in engineering and construction that refers to a list of materials with quantities and types (such as specific grades of steel) that are required to build a designed structure or item. This list is generated by analysis of a drawing or other design documents. The list of required materials for construction is sometimes referred to as the material takeoff list (MTOL).

Mechanical Completion/*Finalización Mecánica***:** Mechanical completion is achieved when the key process components of a stand-alone individual project system, subsystem, or single piece of equipment (i.e., a unit system) within a project facility are fully constructed and complete enough for the unit system alone to mechanically, electrically, pneumatically, and hydraulically start (as the case may be) and operationally perform a test run. Minor work that does not interfere with operation, such as punch list and touch-up work, may still remain.

Milestone/*Hito***:** A significant point or event in the project; checkpoints in a schedule.

Milestone Schedule/*Cronograma de Hitos***:** A summary-level schedule that identifies the major schedule milestones within the project.

Monte Carlo Simulation/*Simulación de Monte Carlo***:** A computer-based statistical program that randomly generates thousands of probable performance outcomes based on probability distributions for cost and schedule on individual tasks. The outcomes are used to generate a probability distribution for the project as a whole, from which a most probable result can be calculated for any selected confidence level.

Net Present Value (NPV)/*Valor Presente Neto (VPN)***:** The difference between the present value of cash inflows and the present value of cash outflows.

Operations Readiness Plan (ORP)/*Plan de Alistamiento Operativo***:** The project document that describes how the project will be phased into the operating business at project completion.

Potential Deviation Notice (PDN)/*Aviso de Desviación Potencial*: A form used in change management control to capture potential changes. PDNs are used for both trends and trends that become change orders.

Preliminary Economic Assessment (PEA)/*Evaluación Económica Preliminar:* Redefined in Canadian National Instrument 43-101 *Standards of Disclosure for Mineral Projects* in August 2012 as an early stage document for public disclosure of the results of conceptual economic studies. A PEA is deliberately kept separate and distinct from the more comprehensive prefeasibility and feasibility studies. A PEA could be a component of a scoping evaluation as defined in this handbook, but in itself, a PEA would not normally be sufficiently encompassing to qualify as a scoping evaluation.

Process Flow Diagram (PFD)/*Diagrama de Flujo de Procesos*: A diagram used in chemical and process engineering to indicate the general flow of both mine and plant processes and equipment. A PFD displays the relationship between the major equipment of a mine plant facility but does not show minor details such as piping details and designations.

Project Charter/ *"Charter" del Proyecto*: As utilized within this Companion Guidebook the project charter defines the work plan for the viability stage PFS and FS. This will clarify what work must be done by the Owner prior to the PFS or FS kickoff along with setting up the project correctly and what are the design develop objectives. It is a work plan that outlines the how, why, who, what, and where for the feasibility study activities.

Project Contingency/ *(Not Translated)*: A specified amount of money for goods and services which, at the current state of project definition, cannot be accurately quantified, but that history and experience show will be necessary to achieve the given project scope.

Project Control Budget/*Presupuesto de Control del Proyecto*: The approved budget for which the project manager is responsible.

Project Controls Manager/*Gerente de Control de Proyectos*: The individual responsible for managing project controls on the project; typically a person with specialist project controls background who reports directly to the project manager.

Project Definition Rating Index (PDRI)/*Indice de Calificación de la Definición del Proyecto:* Introduced by the Construction Industry Institute (CII) as a front-end project-planning aid to measure a project's completeness with regard to concept, feasibility, and scope definition.

Project Description/*Descripción del Proyecto*: A formal detailed written depiction of the entire work scope of the project, along with supporting studies, maps, and drawings that explain the project's goals, deliverables, challenges, resources to be used, execution methods, people and organizations involved, and other relevant information necessary to understand the project being undertaken.

Project Design Basis Document/*Documento de Bases de Diseño del Proyecto*: A formal document that contains all project design criteria, specifications, and quality mandates required of the engineer and construction contractors. It includes sufficient description, data, flow diagrams, and test support to fully capture all facets of the project design.

Project Execution Plan (PEP)/*Plan de Ejecución del Proyecto*: A documented description, in narrative and logic network format, that describes and coordinates the sequence of activities necessary to engineer and execute a project in the field as well as the responsibilities and strategy to timely meet project objectives relating to those activities. It defines responsibilities and sets out the project strategy in relation to the selection and usage of appropriate outside assistance, for example, engineering, procurement, and construction management (EPCM) contractors and others.

Project Governance/*Governanza del Proyecto*: The management organization framework within which project decisions are made to ensure the effectiveness of projects.

Project Management/*Gestion de Proyecto*: In its simplest form, the application of knowledge, skills, tools, processes, and techniques to meet the project objectives. Within *Project Management for Mining*, project management is given a more focused definition: It is the process by which a team of people guide a project using the elements of planning, analyzing, directing, monitoring, problem solving, and communicating to take an idea from the opportunity stage through development to the achievement of specific, established corporate objectives within set cost, schedule, and quality constraints.

Project Risk/*Riesgo del Proyecto:* An unforeseen event or activity that can impact the project's progress, result, or outcome in a negative way.

Project Schedule/*Cronograma del Proyecto*: The planned dates for performing schedule activities and for meeting schedule milestones.

Project Scope/*Alcance del Proyecto*: The work that must be performed to deliver the products, services, and results with the specified features and functions as described in the Project Execution Plan (PEP) and its accompanying approved Authorization for Expenditure (AFE).

Quality Assurance (QA)/*Aseguramiento de Calidad*: Focuses on process input in the quest for reducing the potential for error within a project. QA encompasses the planned and systematic activities that ensure the quality requirements for the project will be fulfilled.

Quality Control (QC)/*Control de Calidad:* Focuses on process output for a project. QC embraces all the techniques and activities that fulfill the project requirements for quality, including observation, checking, and verification.

Quality Management Plan/*Plan de Gestión de Calidad*: The plan that describes how the project management team will deliver the quality requirements of the Owner. The quality management plan and its quality assurance and quality control (QA/QC) elements form a component of the Project Execution Plan (PEP).

Ramp-Up/*(Not Translated)*: Follows the start-up period and takes the operating facility to its full target production design capacity.

RACI chart/*(Not Translated)*: An assignment matrix chart clarifying participation interaction by various project roles when undertaking tasks. The responsibility assignment matrix chart describes the participation by various roles in completing tasks or deliverables for a project. It is especially useful in clarifying roles and responsibilities during execution of project stages. RACI stands for

R = Responsible—the person responsible for delivering the project or task successfully;

A = Accountable—the person with ultimate accountability and authority (R reports to A);

C = Consulted—someone whose input adds value or whose buy-in is needed for implementation; and

I = Informed—persons notified of results or actions taken but are not part of the decision.

Reliability/*Fiabilidad*: Further data development, analysis, engineering, would not significantly change the currently defined project data models, project components, schedule, cost estimates or output production.

Request For Proposal (RFP)/*Solicitud de Propuesta*: A formal document requesting a proposal and cost to perform the services necessary for a defined scope of work (SOW).

Risk Management/*Gestión de Riesgos*: Process of identifying, quantifying, and managing the risks that can prevent an organization from achieving its objectives.

Risk Mitigation/*Mitigación de Riesgos*: A risk response planning technique that seeks to reduce the probability of occurrence or impact of a risk to below an acceptable threshold.

Scope Change/*Cambio de Alcance:* A conceptual change to the Project Execution Plan (PEP) and its accompanying approved Authorization for Expenditure (AFE). A scope change almost always requires an adjustment to the project cost or schedule.

Scope Of Work (SOW)/*Alcance de Trabajo*: A document that describes the physical and temporal boundaries of the project. The SOW defines what the project will and will not deliver. It contains the project goals and tasks, and the work required to accomplish them. SOW is also known as statement of work.

Scoping Evaluation/*Evaluación de Alcance*: A preliminary exercise or research that is meant to define the scope of a project. It is a first-pass, order-of-magnitude estimate of the project, based on the minimal data available, to assess whether the opportunity afforded by the project is potentially economically viable.

Section/*Sección*: Refers to the Companion Guidebook which is divided into sections.

SEC Rule S-K 1300 Technical Report (U.S. Securities and Exchange Commission)/*(Not Translated)*: Regulation S-K 1300 (Items 1300–1305) is a mandatory SEC mandate requiring US-listed mining companies to disclose mineral resources and reserves using CRIRSCO-based standards. Effective since Jan 1, 2021, it requires a "qualified person" (QP) to prepare a Technical Report Summary (TRS) for material properties. The rule ensures standardized reporting for exploration, development, and production-stage properties. The SK1300 Technical Report is deliberately kept separate and distinct from the more comprehensive prefeasibility and bankable (definitive) feasibility studies.

Social Acceptance/*Aceptación Social*: The disposition of the community to tolerate, agree to, or consent to the project. This is generally perceived as the lower level of a social license.

Social Approval/*Aprovación Social*: A favorable regard of, agreeing to, or being pleased with the project. This is sometimes characterized as full social license because it recognizes that the project has earned a higher level of social approval.

Stakeholder/ ***(Not Translated)*****:** Persons and parties who are participants in the project (active or passive) with a legitimate interest (business, legal, social, or ethical) in the outcome of the project. Stakeholders include project proponents, project management, project team, project contractors and consultants, project customers, the project Owner, future project employees, and affected (or potentially affected) members of the local community. In other words, stakeholders are the persons or organizations actively involved in the project or whose interest may be positively or negatively affected by the execution or completion of the project. A stakeholder may also be one who exerts influence over the project and its deliverables.

In the context used within the text rarely, if ever, would the term *stakeholder* include project regulators (i.e., governmental persons) or project neighbors (outside persons not materially affected by the project)—these entities are more "interested bystanders" than project participant

Sustaining Capital/ ***Capital de Sostenimiento*****:** The periodic addition of capital that is required to maintain operations at design levels for the predicted life of the mine.

Time and Materials (T&M)/ ***Tiempo y Materiales*****:** A standard type of contract used in construction in which the buyer agrees to wholly reimburse the contractor for the work performed to complete construction by the contractor's employees and subcontractors, and for the materials used in the construction (plus the contractor's markup). T&M is in contrast to a fixed-price contract in which the buyer agrees to pay the contractor a set lump sum for fulfillment of the contract no matter what the contractor pays his or her employees, subcontractors, and suppliers.

Turnkey/ ***Llave en Mano*****:** Denotes a facility, process, or equipment that is built, supplied, and installed or delivered in a completed state, ready to operate.

Work Breakdown Structure (WBS)/ ***Estructura de Desglose del Trabajo*****:** A deliverable-oriented hierarchical decomposition of the project work to be executed to accomplish the project objectives and create the required deliverables. Within the WBS, the project scope of work (SOW) is broken down into distinct controllable size elements. Each element is assigned a distinct cost. The WBS organizes and defines the total scope of the project.

Zero-Based Estimate/ ***Estimado de Base Cero:*** A bottom up development of costs, starting with a "zero base" for every line item. Each item is justified and analyzed for need and cost. Budget is built solely on what is needed for the given scope. Factoring or comparison costing is not zero based estimating.

Acronynms

AACE	Association for the Advancement of Cost Engineering International
AFE	Authorization for Expenditure: See Glossary
ASPE	American Society of Professional Estimators
BOD	Board of Directors: for the Owner mining company
BOOT	Built-Operate-Own-Transfer contract agreement
C	Construction
CAD	Computer-Aided Design or Computer-Aided Drafting
CCC	Care Custody & Control: term for when facility responsibility shifts from contractor to Owner
CEO	Chief Executive Officer
CG	The *Companion Guidebook* to the *Project Management for Mining* 2nd edition.
CII	An acronym for Construction Industry Institute
CM	Construction Manager
CO	Change Order
Constr.	Construction
CP	Critical Path
CPI	Cost Performance Index: Earned Value measure of cost efficiency on a project
CPM	Critical Path Methodology: scheduling process to determine activities critical to project completion
CRIRSCO	Committee for Mineral Reserves International Reporting Standards: oversight of mineral reserves reporting
DD	Diamond Drilling: drilling method where diamond bit is used to core out rock and return through the non-rotating inner tube using mud or water to lubricate and cool the system.
DOE	Department of Energy: of US Government
E	Engineering
E & P	Engineering and Procurement

EA	Environmental Assessment: a 1st order environmental review in US, 2nd order review in Canada
EIS	Environmental Impact Statement: a 2nd order environmental review in US
EMP	Environmental Management Plan
Engr.	Engineering
EPC	Engineering Procurement & Construction: See Glossary
EPCM	Engineering, Procurement, and Construction Management: See Glossary
ERP	Enterprise Resource Planning: See Glossary
ESIA	Environmental & Social Impact Assessment: 2nd order environmental review in Europe and for World Bank
ESS	Environmental Impact, Permits, Social Acceptance & Sustainability
Est.	Estimate
EV	Earned Value : work performed expressed in terms of the approved budget.
EVM	Earned Value Management: project management technique for objectively measuring project performance.
Exec.	Executive
FEED	Front-End Engineering Design: an alternative name for FEL. FEED depth is similar to "basic engineering".
FEL	Front End Loading; robust planning and design early in project life when ability to influence change is still high.
FS	Feasibility Study
GA	General Arrangement: See Glossary
GAP Analysis	Technique to determine steps to take, or missing items to add, to move from current status to target state
GM	General Manager
GMP	Guaranteed Maximum Price: See Glossary
Govt.	Government
HAZOP	Hazard & Operability Review: See Glossary
HR	Human Resources
IFC	International Finance Corporation: institution that encourages private-sector growth in developing nations.
IFC	Issued for Construction: drawings ready for use by constructors in the field.
IOM	Institute of Materials, Minerals and Mining
IPA	Independent Project Analysis: a firm that researches projects to help clients improve use of capital assets.
ISO	International Organization for Standardization
IT	Information Technology

JORC	Joint Ore Reserves Committee: Australasian Code for Reporting of Exploration Results.
JV	Joint Venture
KPI	Key Performance Indicator: an indicator measurement of the health/effectiveness of the process/function.
LD	Liquidated Damages: compensation set upfront in contract for injured party to collect upon a specific breach.
LE	Lead Engineer
LS	Lump Sum: See Glossary
LTA	Lost Time Accident
Mgmt.	Management
MSHA	Mine Safety and Health Administration
NGO	Nongovernmental Organization
NI 43-101	National Instrument 43-101: Canadian mineral resource classification scheme used for mineral properties.
NI 43-101 PEA	NI 43-101 Preliminary Economic Assessment: a preliminary report that includes inferred mineral resources too speculative geologically to have economic considerations applied for categorization as a mineral reserve.
NPV	Net Present Value See Glossary
O&G	Oil and Gas
O&M	Operations and Maintenance
O&M Manuals	Operating and Maintenance Manuals: instructions for personnel tasked to make installed equipment function.
OEM	Original Equipment Manufacturer: producer of a complete end product, or sub-assembly used in end product.
OPEX	An acronym for operating (or operations) expenditures.
Ops.	Operations
Org. Chart	Organization Chart
ORP	Operations Readiness Plan: describes how the project will be phased into operations at project completion.
OSHA	Occupational Safety and Health Administration
P90	Denotes a 90% probability of an event happening.
P&L Statement	Profit and Loss Statement: financial statement summarizing revenues and expenses in a specific time period.
P&ID	Piping & Instrumentation Diagram: shows interconnection of process equipment and instrumentation utilized.
PCS	Project Control System: monitors and reports aspects of project execution to original SOW, budget & schedule.

PDN	Potential Deviation Notice: the capture on paper of a potential deviation to original project plan.
PE	Professional Engineer: each US state regulates the practice of engineering by granting PE licenses.
PEA	Preliminary Economic Analysis: See Glossary
PEP	Project Execution Plan: See Glossary
PFD	Process Flow Diagram: See Glossary
PM	Project Management
PM	Project Manager
PMfM	Project Management for Mining: *Project Management for Mining* 2nd edition
PMI	Project Management Institute
PO	Purchase Order
PPP	Pre-Project Planning: alternative name for FEL or FEED.
PPM	Project Procedures Manual: document captures all project policies, procedures & requirements in one place
Prelim.	Preliminary
PV	Planned Value: the authorized budget assigned to a scheduled work activity or WBS component.
PWC	PricewaterhouseCoopers: a multinational professional services and auditing firm, headquartered in UK.
QA	Quality Assurance See Glossary
QC	Quality Control See Glossary
RACI	Responsible, Accountable, Consulted, Informed
RC	Reverse Circulation: drilling method that injects compressed air into the drill pipe to collect rock samples
RFI	Request for information.
RFP	Request for Proposal: See Glossary
RFQ	Request for Qualifications: request to gauge whether contractors possess necessary qualifications to bid
RFQ	Request for Quotation: request for cost of supply for a clearly specified item of equipment or material
SEC Form 10-K	US Securities & Exchange Commission Form 10-K: Annual required report summarizing company performance
SOP	Standard Operating Procedures: set of instructions to help workers carry out complex routine operations
SOW	Scope of Work: document that describes and defines the physical and temporal boundaries of the project
Specs.	Specification sheets: documented requirements to be satisfied by material, design, product, or service.

SPI	Schedule Performance Index: Earned Value measure of schedule efficiency on a project.
STD DEV	Standard Deviation: statistical term quantifying the spread of numbers in a set of data from its mean value
SWAT Team	Highly trained, experienced team who tackle situations beyond the capabilities of the assigned personnel.
T&Cs	Terms and Conditions: standard arrangements, provisions, requirements & rules that are part of a contract
T&M	Time and Materials: See Glossary
TOR	Terms of Reference: describes the purpose, scope and structure of the project, for regulatory authorities
VAT	Value-added Tax: also known in some countries as a goods & services tax (GST), is a general consumption tax.
VP	Vice President
WBS	Work Breakdown Structure: See Glossary

www.ingramcontent.com/pod-product-compliance
Lightning Source LLC
LaVergne TN
LVHW070107021026
849160LV00026B/1699
* 9 7 8 0 8 7 3 3 5 0 1 5 0 *